Dietary Supplements in Human Health and Disease

Dietary Supplements in Human Health and Disease

Guest Editors

Elisa Benetti
Valentina Boscaro

MDPI

Basel • Beijing • Wuhan • Barcelona • Belgrade • Novi Sad • Cluj • Manchester

Guest Editors
Elisa Benetti
Dipartimento di Scienza e
Tecnologia del Farmaco
University of Turin
Turin
Italy

Valentina Boscaro
Dipartimento di Scienza e
Tecnologia del Farmaco
University of Turin
Turin
Italy

Editorial Office
MDPI AG
Grosspeteranlage 5
4052 Basel, Switzerland

This is a reprint of the Special Issue, published open access by the journal *Nutrients* (ISSN 2072-6643), freely accessible at: https://www.mdpi.com/journal/nutrients/special_issues/U51RW2D034.

For citation purposes, cite each article independently as indicated on the article page online and as indicated below:

Lastname, A.A.; Lastname, B.B. Article Title. *Journal Name* **Year**, *Volume Number*, Page Range.

ISBN 978-3-7258-6694-6 (Hbk)
ISBN 978-3-7258-6695-3 (PDF)
https://doi.org/10.3390/books978-3-7258-6695-3

Contents

About the Editors

Elisa Benetti

Elisa Benetti is an Associate Professor in Pharmacology at the Dept. of Drug Science and Technology, University of Turin. She is a co-author of more than 40 research papers. Her main research field is the study of new pharmacological strategies for the treatment of type 2 diabetes, with a particular focus on insulin resistance and its link with inflammation. She is a member of the European Association for the Study of Diabetes (EASD) and the Italian Society of Pharmacology (SIF).

Valentina Boscaro

Valentina Boscaro is an Assistant Professor of Pharmacology at the Dept. of Drug Science and Technology, University of Turin. Her main research interests focus on the investigation of NLRP3 inhibitors in in vitro cellular models and on the antiproliferative activity of novel compounds in both tumor and non-tumor cell lines. More recently, her research has expanded to include the study of substances with neuroprotective effects. She is a member of the Italian Society of Pharmacology (SIF).

Preface

In recent years, interest in dietary supplements has grown substantially, driven by the increasing prevalence of lifestyle-related chronic conditions and the search for complementary strategies to support health and well-being. Despite their widespread use, many supplements remain characterized by incomplete evidence, unclear mechanisms of action, and limited safety evaluation. This awareness motivated the development of a Special Issue dedicated to a rigorous and multidimensional examination of natural compounds with emerging or established relevance in human health.

The contributions gathered in this Reprint reflect the diversity and complexity of the field.

We hope that this collection will serve as a valuable resource for researchers, clinicians, and professionals in the field, fostering new insights and promoting a more evidence-based and informed use of dietary supplements.

Elisa Benetti and Valentina Boscaro
Guest Editors

nutrients MDPI

Article

Styphnolobium japonicum Fruit and Germinated Soybean Embryo Complex Extract for Postmenopausal-Symptom Relief

Jeong-Won Ahn [1], Hyun-Soo Kim [1], Kongara Damodar [2,3], Hee-Hyun Shin [4], Kyung-Mi Kim [4], Jung-Youl Park [5], Su-Kil Jang [1,3], Yeong-Min Yoo [2], Jae-Chul Jung [4] and Seong-Soo Joo [1,3,*]

1 Department of Marine Bioscience, College of Life Science, Gangneung-Wonju National University, Gangneung 25457, Gangwon, Republic of Korea; 0000@gwnu.ac.kr (J.-W.A.)
2 East Coast Life Sciences Institute, College of Life Science, Gangneung-Wonju National University, Gangneung 25457, Gangwon, Republic of Korea
3 Huscion MAJIC R&D Center, 331 Pangyo-ro, Seongnam 13488, Gyeonggi, Republic of Korea
4 Life Science Research Institute, NOVAREX Co., Ltd., Cheongju 28220, Chungbuk, Republic of Korea
5 Glocal University Project Group, Andong National University, 1375 Gyeongdong-ro, Andong 36729, Gyeongbuk, Republic of Korea
* Correspondence: ssj66@gwnu.ac.kr; Tel.: +82-33-640-2856

Abstract: Background/Objectives: Hormonal alterations during menopause result in substantial physiological changes. Although hormone replacement therapy (HRT) is widely used as a treatment strategy for these changes, its use remains controversial due to its associated risks. Plant isoflavones are phytoestrogens that are considered a potential alternative therapy for postmenopausal syndrome. We aimed to investigate the efficacy of ethanolic extracts from *Styphnolobium japonicum* fruit (SJF) and germinated soybean embryo (GSE) in alleviating prominent menopausal symptoms. Methods: A cell model (MCF7 human breast cancer cells) was used to investigate estrogen-like activity. A rat ovariectomy model was used to simulate estrogen depletion after menopause and to evaluate the efficacy of the SJF–GSE complex extract at ratios of 1:1, 1:2, and 2:1. Results: Treatment with the SJF–GSE extract elicited estrogen-like effects, raising pS2 and estrogen receptor α expression in MCF7 cells. The extract was found to contain 48–72 mg/g sophoricoside and 8–12 mg/g soyasaponin 1, identified as active compounds. In ovariectomized rats, the extract effectively reduced body weight and fat content, alleviated vasomotor symptoms, improved vaginal mucosal health, and exerted osteoprotective effects by enhancing bone density and structure, reducing bone-resorption markers and positively altering estradiol levels and lipid profiles. Conclusions: The SJF–GSE extract, working synergistically, provides a safe and effective alternative to HRT for managing postmenopausal symptoms and enhancing bone health, without adverse effects. These findings support the inclusion of SJF and GSE in health-functional foods and underscore the importance of further research into plant-based therapies for menopause.

Keywords: menopause; *Styphnolobium japonicum* fruit; germinated soybean embryo; ovariectomized rat model; phytoestrogen; medicinal plant

1. Introduction

Menopause, a crucial transition in a woman's life, is marked by significant hormonal changes that lead to a range of physiological and psychological symptoms. These symptoms, including hot flashes, mood swings, osteoporosis, and changes in metabolic health, can severely affect quality of life [1]. While hormone replacement therapy (HRT) is traditionally used to manage these symptoms, its associated risks have propelled the search for safer natural alternatives [2]. Phytoestrogens, plant-derived compounds found in foods, such as soybeans, flaxseed, and whole grains, mimic the effects of estrogen and present a safer option than traditional HRT [3]. Their capacity to bind to estrogen receptors (ERs) and potentially alleviate menopausal symptoms has attracted significant attention from the scientific community, largely because they present fewer risks than conventional HRT [4].

In the realm of herbal medicine, combining specific plants has shown promise in synergistically alleviating postmenopausal symptoms. Previous studies suggest that such combinations may enhance therapeutic outcomes by interacting in ways that complement and amplify their individual effects, thereby offering a holistic approach to menopause management [5,6]. Germinated soybean embryo, derived from soybeans shortly after the onset of germination but before full sprouting occurs, contains high amounts of bioactive compounds such as isoflavones and soyasaponins [7]. These compounds have been shown to mitigate hormonal imbalances and improve bone health in postmenopausal women [8,9]. Specifically, soyasaponin 1 supports bone health and possesses anti-inflammatory, anti-cancer, anti-obesity, and antioxidant properties [10,11]. *Styphnolobium japonicum* L. (synonym *Sophora japonica* L.) fruit (known as Fructus sophorae) is rich in flavonoids (including sophoricoside) and is known for its antioxidative benefits and hormone-regulatory potential [12]. This study, therefore, focuses on these compounds to better understand their contributions to the observed biological effects.

Here, we aim to elucidate the therapeutic potential of combining ethanol extracts of *S. japonicum* fruit and germinated soybean embryo in an ovariectomized rat model that simulates postmenopausal estrogen depletion. While focusing on the ratio of the extracts in this combination, we examine its efficacy in addressing key menopausal challenges, including vasomotor symptoms, vaginal health, bone integrity, and hormonal balance, positioning this natural solution as a viable and safe alternative to traditional HRT. These findings contribute to the burgeoning field of natural plant-based therapies, offering new insights and potential pathways for improving postmenopausal health.

2. Materials and Methods

2.1. Sample Preparation

The *S. japonicum* fruit (SJF) and germinated soybean embryo (GSE) complex extracts were provided by NOVAREX Co., Ltd. (Osong, Chungbuk, Republic of Korea). *S. japonicum* fruit extract was derived from the dried ripe fruits of *S. japonicum* L., and germinated soybean embryo extract was prepared from soybeans after controlled germination. Both extracts were obtained using 60 ± 10% food-grade ethanol (Product No. 0019-200L, CAS No. 64-17-5; Korea Ethanol Supplies Company, Seoul, Republic of Korea) via reflux extraction for 4–6 h at 70 ± 10 °C to concentrate key bioactive components, including sophoricoside and soyasaponin 1. The concentrated extracts were filtered, spray-dried to produce the final SJF extract (SJFE) and GSE extract (GSEE) powders, and stored at room temperature. The extracts were dissolved in dimethyl sulfoxide to a concentration of 100 mg/mL and used as a stock solution. To determine the optimal ratio of these extracts, the physiological activity of each extract and the levels of their active compounds were assessed in vitro, at ratios of 1:1, 1:2, and 2:1. For in vivo studies, the final complex extract, containing concentrations of 48–72 mg/g sophoricoside and 8–12 mg/g soyasaponin 1, was prepared by blending SJF and GSE at the identified optimal ratio and yield. A ratio of 1.5:1 (SJF:GSE) satisfied the established criteria for industrial-scale production.

2.2. HPLC Analysis

Levels of sophoricoside and soyasaponin 1, the active components of the extracts, were analyzed via 1260 Infinity HPLC using a UV detector (Agilent Technologies, Santa Clara, CA, USA). Chromatographic separation of sophoricoside was performed on a reverse-phase HPLC column (YMC-Triart C18, 4.6 mm × 250 mm, S-5 μm, 12 nm; YMC, Kyoto, Japan) at 30 °C, with a mobile phase comprising 0.1% acetic acid in water (A) and 0.1% acetic acid in acetonitrile (B) at a flow rate of 1 mL/min. The gradient program was as follows: 0–5 min, 15% B; 5–15 min, 15–45% B; 15–17 min, 45–50% B; 17–21 min, 50% B; 21–23 min, 50–15% B; and 23–25 min, 15% B. The injection volume was 5 μL and the detection wavelength was 260 nm.

Chromatographic separation of soyasaponin 1 was performed using a reverse-phase HPLC column (YMC-Triart C18, 4.6 mm × 250 mm, S-5 μm, 12 nm; YMC) at 30 °C. The mobile phase was 0.05% formic acid in distilled water and acetonitrile (at 6:4, *v*/*v*), flowing at 1 mL/min for 25 min under isocratic conditions. The injection volume was 10 μL and the detection wavelength was 210 nm.

2.3. Cell Culture

Three distinct cell lines were obtained from the Korean Cell Line Bank (KCLB, Seoul, Republic of Korea) and used to assess the impact of SJFE and GSEE. The human breast cancer cell line MCF7 was cultured in Roswell Park Memorial Institute (RPMI)-1640 medium (Hyclone, Logan, UT, USA), supplemented with 10% fetal bovine serum (FBS; Hyclone), 100 U/mL of penicillin, and 100 μg/mL of streptomycin (Thermo Fisher Scientific, Waltham, MA, USA). Cells were cultured at 37 °C in a humidified atmosphere containing 5% CO_2. For gene analysis, the cells were seeded at 5.0×10^5 cells/well in 12-well plates and allowed to stabilize for 12 h before being treated with 100 nM 17β-estradiol (E2, Sigma-Aldrich, St. Louis, MO, USA) and 200 μg/mL of each test substance in serum-free medium. Additionally, RAW 264.7 (mouse macrophage) and MG63 (human osteoblast-like) cell lines were used to evaluate nitric oxide (NO) production and overall cellular activity. These cells were maintained in Dulbecco's Modified Eagle's Medium (DMEM, Hyclone) supplemented with 10% FBS and antibiotics.

2.4. Cell Viability Assay

To evaluate biological activity, cell viability was determined using the Ez-Cytox Enhanced Cell Viability Assay Kit (WST-8; DoGenBio, Seoul, Republic of Korea), according to the manufacturer's instructions. Briefly, cells were seeded in 96-well plates at a density of 5.0×10^4 cells per well in serum-free DMEM and were treated with various concentrations of the sample. Following treatment for 24 h, the assay reagent was added to each well and the plates were incubated for an additional 2 h at 37 °C. Color intensity, which is directly proportional to the number of viable cells, was quantified by measuring absorbance at 450 nm using a microplate reader (SpectraMax 340, Molecular Devices, San Jose, CA, USA). Viability was expressed as a percentage of the viability in the control group, which was considered to be 100%.

2.5. NO and DPPH Assay

We used a mouse macrophage cell line (RAW 264.7) to assess NO production, an essential marker of the inflammatory response, in response to varying concentrations of SJFE, GSEE, and the complex extract. Cells were plated at a density of 1.0×10^5 cells/well in a 96-well plate and allowed to adhere before replacing the medium with 100 μL fresh serum-free medium in each well. The cells were then treated with the test substances at concentrations ranging from 10 to 100 μg/mL, followed by incubation for 1 h. To induce inflammation, 1 μg/mL of lipopolysaccharide (LPS, O55:B5, Sigma-Aldrich) was added, and after 20 h of incubation at 37 °C in a 5% CO_2 environment, NO production was quantified. This was achieved using the Griess reagent system (Sigma-Aldrich) following the manufacturer's instructions. A nitrite standard calibration curve (Promega, Madison, WI, USA) was used for quantification. A DPPH (2,2-diphenyl-1-picrylhydrazyl, Sigma-Aldrich) radical scavenging activity assay was performed to screen the antioxidant activity of SJFE and GSEE. In each well of a 96-well plate, 10 μL of each extract was combined with 90 μL of 0.2 mM DPPH solution in methanol. After 10 min of incubation at room temperature, absorbance was measured at 517 nm using a microplate reader. DPPH scavenging activity was calculated using the following formula: scavenging activity (%) = ((control absorbance − sample absorbance)/control absorbance) × 100.

2.6. MG63 Activation and RANKL Gene Expression Analysis

We used MG63 osteoblast-like cells and RAW 264.7 macrophage cells to evaluate the modulation of receptor activator of nuclear factor κB ligand (RANK ligand, RANKL) gene expression, a key factor in bone metabolism. The RAW 264.7 cells were initially treated with 1 μg/mL of LPS for 24 h, resulting in a >40-fold increase in NO production relative to the untreated control. This significant increase in NO levels indicates successful macrophage activation. Subsequently, MG63 cells were seeded at a density of 3.0×10^5 cells/well in 12-well plates and allowed to stabilize for 12 h. After stabilization, the cells were cultured in fresh serum-free media and treated with varying concentrations of the test substances, ranging from 10 to 250 μg/mL. One-hour post-treatment, the MG63 cells were exposed to an NO-rich medium derived from activated RAW 264.7 cells, with the NO concentration adjusted to 50 μM. Following a 20-h incubation period in the NO-rich environment, we extracted total RNA from MG63 cells and analyzed gene expression.

2.7. qPCR Assay

To analyze changes in gene expression, total RNA was extracted from selected cells or tissues using TRIzol reagent (Thermo Fisher Scientific), following the manufacturer's guidelines. To synthesize complementary DNA (cDNA), 1.5 μg of total RNA was used as a template. cDNA synthesis was performed using the ImProm-II Reverse Transcriptase System (Promega), following the manufacturer's protocol. The qPCR reactions were prepared using SensiMix SYBR Hi-ROX PCR Master Mix (Bioline, London, UK) and performed as previously described [13]. The final results are presented in terms of fold change relative to the sham group, providing a clear comparison of gene expression across different experimental conditions. The primer sequences used are listed in Table S1.

2.8. Western Blot Analysis

To assess estrogen-like activity in MCF7 cells, the cells were plated at 1.0×10^6 cells/well in 60 mm culture dishes and allowed to stabilize for 12 h. The cells were treated with 100 nM E2 and 200 μg/mL of each test substance in a serum-free medium for 3 h. Total protein from the cells was extracted using radioimmunoprecipitation assay (RIPA) buffer supplemented with a protease and phosphatase inhibitor cocktail (Roche, Basel, Switzerland). Western blotting was performed as described previously [13]. Protein bands were visualized using an enhanced chemiluminescence (ECL) solution (Thermo Fisher Scientific), and luminescence signals were captured using a chemiluminescence imaging system (LuminiGraph II; ATTO, Tokyo, Japan) and quantitatively analyzed using ImageJ software (Version 1.51j; NIH, Bethesda, MD, USA). The results are expressed as relative protein expression and presented as fold-change relative to the control group. The antibodies used are listed in Table S2.

2.9. Immunocytochemical Analysis of ERα in MCF7 Cells

MCF7 cells were seeded in four-well chamber slides (1.0×10^5 cells/well) and stabilized for 12 h. After stabilization, the cells were incubated in fresh serum-free medium with 100 nM E2, 200 μg/mL of the complex extract, or 1 μM ER antagonist (ICI 182,780, Sigma-Aldrich) for 24 h. The cells were then fixed in 4% paraformaldehyde (30 min), permeabilized with 0.25% Triton X-100 (20 min), and blocked with 1% BSA to reduce non-specific antibody binding (1 h). Afterward, the cells were incubated with anti-ERα primary antibody (sc-7207, Santa Cruz Biotechnology, Santa Cruz, CA, USA; 1:200 dilution) overnight at 4 °C, followed by incubation with Alexa Fluor 488-conjugated anti-rabbit IgG secondary antibody (Cell Signaling Technology, Danvers, MA, USA; 1:500 dilution) for 1 h at room temperature. Propidium iodide (PI) was used for nuclear counterstaining (20 min at room temperature). Between each step, the slides were washed with phosphate-buffered saline three times after each major step and five times after counterstaining. ERα expression and distribution were analyzed using fluorescence microscopy at 600× magnification (Eclipse Ti-S; Nikon, Tokyo, Japan).

2.10. Animal Housing and Ovariectomy

We used female Sprague-Dawley (SD) rats, aged 12 weeks and weighing 220–240 g (KOATECH, Gyeonggi, Republic of Korea). The animals were housed in a controlled environment at 21 ± 2 °C, 40–60% humidity, and a 12-h light/dark cycle. Before starting the experiment, the rats underwent a week-long acclimatization period. The rats were divided into six groups (Figure 1): (1) sham-operated (non-ovariectomized control, Sham); (2) ovariectomized control (OVX); (3) OVX treated with 0.1 mg/kg of E2 (positive control); groups 4–6 were OVX groups that received the complex extract at 25 mg/kg (Low), 50 mg/kg (Medium), and 100 mg/kg (High), respectively. The ovariectomy procedure involves dorsal fur shaving, dorsal incision, bilateral ovary removal, and suturing [14]. The Sham group underwent the same surgical procedure, but without ovary removal. Postoperative care was diligently administered to ensure seamless recovery of the animals, without complications, such as incision rupture or inflammation. Only fully recovered animals advanced to the subsequent phase of the experiment (n = 7). To establish a reliable OVX rat model for efficacy studies, estrogen depletion was confirmed following ovariectomy. During the 3-week recovery period, the experimental animals were supplied with AIN-76A phytoestrogen-free diet (Harlan Teklad, Madison, WI, USA). The rats were then fed diets formulated according to their group for 12 weeks. Specifically, the Sham and OVX control groups were provided ad libitum diets AIN-76A, whereas the treatment groups were provided with custom-manufactured diets infused with the specified treatment compounds (Daehan Biolink, Chungbuk, Republic of Korea). Body weight was measured twice per week at regular intervals. All animal experiments were approved by the Institutional Animal Care and Use Committee (IACUC) of Gangneung-Wonju National University, Gangneung, Republic of Korea (GWNU-2020-18), and were conducted following the relevant guidelines and regulations and ARRIVE guidelines. Stress was minimized by providing feed mixed with the extracts, and consistent conditions were maintained. No abnormalities were observed, indicating that the changes were due to the treatment.

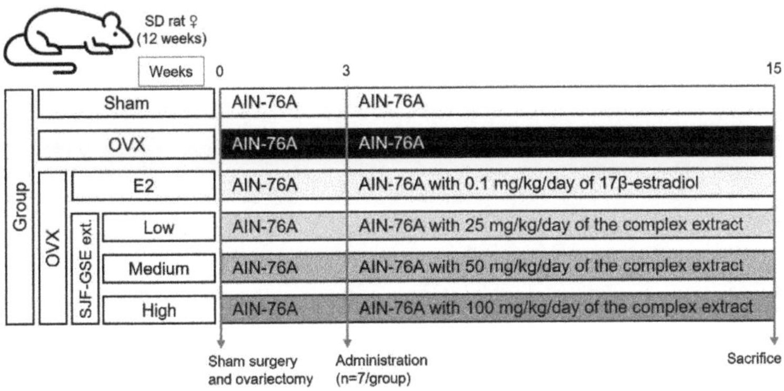

Figure 1. In vivo study design and group allocation. The flowchart illustrates the design of our in vivo study involving adult female Sprague-Dawley (SD) rats. The rats were divided into six groups (n = 7 per group): (1) non-ovariectomized control (Sham); (2) ovariectomized control (OVX); (3) positive control group receiving 0.1 mg/kg of 17β-estradiol (E2); groups 4–6 were OVX groups treated with the blended SJF–GSE complex extract at 25 mg/kg (Low), 50 mg/kg (Medium), and 100 mg/kg (High), respectively. After the 3-week recovery period, the rats were provided with custom-manufactured diets containing the specific treatment compounds. The SJF-GSE complex extract was prepared as detailed in the Section 2, with a 1.5:1 ratio of SJF to GSE. SJF, *Styphnolobium japonicum* fruit; GSE, germinated soybean embryo.

2.11. Measurement of Vasomotor Symptoms

Vasomotor symptoms, which typically appear as initial indicators of menopause in women, were assessed in our postmenopausal rat model. To accurately monitor and evaluate these vasomotor symptoms, we meticulously measured the rectal temperatures of the rats 2 d before their sacrifice. The temperatures were measured at 10-min intervals over a span of 120 min, immediately after 15 min of forced-running on a motorized treadmill (MK-680S, Muromachi Kikai, Tokyo, Japan) at a constant speed of 15 m/min. For each rat, the baseline temperature was established as the average temperature measured 20 min before the forced-running session. Changes in rectal temperature after exercise were recorded for 120 min. To ensure precision and minimize variability in our measurements, rectal temperatures were recorded using a highly accurate electronic thermometer (MT200; Microlife, Taipei, Taiwan). This approach allows for the comprehensive and accurate assessment of vasomotor symptoms, providing crucial insights into the physiological changes associated with menopause in the OVX rat model.

2.12. *Evaluation of Vaginal Epithelial Cell Changes (Vaginal Cornification)*

Vaginal dryness is one of the most common symptoms of menopause. Therefore, we examined changes in the vaginal epithelial cells of each group to assess the efficacy of the extracts in alleviating menopausal symptoms. Vaginal epithelial cells were collected from the vaginal walls of the rats in each group using a sterile swab, immediately before sacrifice. These cells were spread onto a glass slide and stained with Giemsa stain (Sigma-Aldrich). This allowed us to count the number of nucleated and cornified cells to determine the proportion of cornified epithelial cells. A higher proportion of cornified epithelial cells in the vaginal mucosa indicates less vaginal cornification, suggesting that the vaginal mucosa is more flexible and less affected by menopausal dryness.

2.13. *Assessment of Organ Abnormality and Fat Tissues*

Comprehensive organ analyses were conducted to assess the safety of the complex. The process began with a 12-h fasting period, followed by sacrifice via anesthesia using ether. During autopsy, key tissues, such as the liver, spleen, kidneys, stomach, uterus, perirenal fat, and the femurs, were meticulously excised and weighed to detect potential abnormalities or changes. To examine body fat accumulation, histological analysis of adipose tissue was performed. Adipose tissues were fixed in 10% neutral buffered formalin (NBF) for 24 h. The fixed tissues were dehydrated through a graded series of alcohol solutions, cleared in xylene, and infiltrated with paraffin wax. Thin sections (5 μm thickness) were cut from the paraffin-embedded tissue blocks and stained with hematoxylin and eosin (H&E) according to conventional procedures. Lipid droplet area was measured using ImageJ software.

2.14. *Blood Hormone and Biochemical Analysis*

We conducted biochemical analyses to evaluate serum markers indicative of metabolic and liver function in the OVX rat model. Arterial blood was collected for hormone and biochemical analyses, and serum samples were stored at −80 °C. Serum E2 was quantitatively measured using an enzyme-linked immunosorbent assay (ELISA) kit (Cat. No. 501890; Cayman Chemical, Ann Arbor, MI, USA). Serum C-terminal telopeptide (CTx) was quantified using the Rat CTX-1 ELISA kit (NBP2-76634; Novus Biologicals, Littleton, CO, USA). Serum osteocalcin was quantified using a Rat Osteocalcin ELISA kit (NBP2-68153; Novus Biologicals). Biochemical parameters, including alanine aminotransferase (ALT), triglycerides (TG), high-density lipoprotein (HDL), low-density lipoprotein (LDL), lactate dehydrogenase (LDH), calcium, and phosphorus levels, were analyzed using an INTEGRA 400 automatic analyzer (Roche, Mannheim, Germany).

2.15. *Microscopic Assessment of Trabecular Bone Loss*

Microscopic evaluation of trabecular bone loss was performed on the femur. Right femurs were fixed in 10% NBF for 24 h and decalcified using 10% ethylenediaminetetraacetic acid (EDTA) at 4 °C for 14 d. Following decalcification, the bones were embedded in paraffin. From these, sections of 5 μm thickness were cut for detailed examination. The sections were then stained with H&E, a technique commonly used to highlight bone microarchitecture. The resulting stained sections provide a clear visualization of the structural integrity of the bone. The area of bone loss was measured using ImageJ software.

2.16. Isolation of Bone Marrow Cells from Femurs

Comprehensive gene expression analyses were performed using the femurs of the OVX rats to elucidate the molecular mechanisms associated with estrogen depletion. After the 12-week treatment period, the rats were euthanized and tissues were collected for analysis. Bone marrow cells were isolated from the left femur using a previously described method, with minor modifications [15]. Briefly, the bones were bisected and placed in microcentrifuge tubes. Bone marrow was harvested via centrifugation at $5000\times g$ for 5 min. The cells were then filtered through a 40 μm cell strainer to remove debris, and erythrocytes were lysed using RBC lysis buffer (BioLegend, San Diego, CA, USA). qPCR was used to analyze the expression of genes pertinent to bone tissue metabolism and health.

2.17. Assessment of Bone Mineral Contents, Bone Mineral Density, and X-ray Imaging

We evaluated the effects of the complex extract on postmenopausal bone health in the OVX rat model, using dual-energy X-ray absorptiometry (DXA) to determine bone mineral content (BMC) and bone mineral density (BMD) measurement, and X-ray imaging for structural observation (PIXImus II densitometer, GE-Lunar, Madison, WI, USA). The left femur was prepared and analyzed via DXA to quantify both BMC and BMD to assess bone integrity. Preparation involved removing the surrounding muscle and tissue and preserving the bone in sterile saline for accurate measurement. X-ray imaging of the femur revealed structural changes, complementing the quantitative and qualitative analyses and providing a holistic view of bone health.

2.18. Statistical Analysis

The results are presented as means ± standard deviation. The significance of differences between experimental groups was analyzed using Student's t-tests or one-way ANOVA followed by Tukey's post-hoc test, using GraphPad Prism Version 5.01 (GraphPad Software Inc., San Diego, CA, USA). Differences were considered significant at $p < 0.05$.

3. Results and Discussion

3.1. Optimal Ratio of SJFE and GSEE

In examining the optimal ratio of SJFE and GSEE, we focused on their antioxidative and anti-inflammatory properties and their effects on osteoclastogenesis-related gene expression. Menopause, marked by a significant decrease in estrogen levels due to the cessation of ovarian follicle function, leads to increased oxidative stress within the body. This elevation in oxidative stress not only exacerbates inflammation but also contributes to obesity, which in turn can further enhance oxidative stress, creating a detrimental cycle of metabolic disturbance [16,17]. These molecular alterations have been implicated in the pathogenesis of postmenopausal osteoporosis [18]. At the concentrations used here, the extracts did not affect cell activity (Figure S1). We first examined the extracts' antioxidant potential using a DPPH scavenging activity assay. This revealed that SJFE exhibited DPPH scavenging activity at a half-maximal inhibitory concentration (IC_{50}) of 61.4 ± 11.2 μg, whereas GSEE demonstrated this activity at an IC_{50} of 293.7 ± 15.7 μg (Table 1; Figure S2).

The extracts' anti-inflammatory potential was assessed by measuring the inhibition of NO production in RAW 264.7 macrophage cells. SJFE and GSEE had median effective dose (ED_{50}) values of 206.7 ± 18.1 μg/mL and 34.2 ± 5.7 μg/mL, respectively (Table 1; Figure S3). The optimal concentrations were selected based on their efficacy and non-toxicity, as indicated by the IC_{50} and ED_{50} values, which represent the concentrations at which 50% of the maximal effect was achieved.

The effectiveness of these extracts in modulating bone metabolism was evaluated by examining their effects on RANKL gene expression in MG63 osteoblast-like cells. The inhibitory effects on RANKL expression were significant ($p < 0.001$) at 250 μg/mL of SJFE and 10 μg/mL of GSEE (Table 1; Figure S4). This indicates that SJFE exhibited greater antioxidant activity than GSEE, while GSEE demonstrated superior anti-inflammatory and RANKL-inhibitory activity.

Table 1. Antioxidant, anti-inflammatory, and osteoprotective activity of *Styphnolobium japonicum* fruit extract (SJFE), germinated soybean embryo extract (GSEE), and their combinations.

Activity	SJFE	GSEE	SJFE and GSEE Combinations		
			1:1	1:2	2:1
DPPH scavenging (IC_{50}, μg)	61.42 ± 11.23	293.75 ± 15.71	117.56 ± 12.04	84.48 ± 13.74	77.72 ± 14.13
NO inhibition (ED_{50}, μg/mL)	206.71 ± 18.11	34.22 ± 5.71	64.82 ± 1.95	48.19 ± 1.88	59.37 ± 1.44
RANKL inhibition ($p < 0.001$, μg/mL)	250	10	100	10	10

The values represent the means ± standard deviation. These results highlight the efficacy of the extracts in scavenging DPPH radicals, inhibiting NO production, and inhibiting RANKL—key indicators of antioxidant, anti-inflammatory, and osteoprotective effects, respectively. For further details, see Figures S2–S4. DPPH, 2,2-diphenyl-1-picrylhydrazyl; NO, nitric oxide; RANKL, receptor activator of nuclear factor κB (RANK) ligand.

In testing antioxidant activity via the DPPH assay, ascorbic acid treatment was used as the positive control; in contrast, anti-inflammatory effects, including inhibition of NO production and of RANKL expression, were evaluated relative to negative controls. Based on these findings, we hypothesized that the two extracts, with their distinct antioxidative and anti-inflammatory properties, could have complementary effects in alleviating menopausal symptoms. Consequently, we further examined the optimal ratio of SJFE and GSEE and their synergistic effectiveness.

The ratio of SJFE and GSEE that maximizes their synergistic effects was examined. Three ratios of SJFE to GSEE (1:1, 1:2, and 2:1) were tested and their DPPH scavenging, NO production inhibition, and RANKL expression inhibition activity were recorded (Table 1). Based on these assays, a 2:1 ratio of SJFE to GSEE was the most effective, balancing the physiological activity of both extracts.

3.2. Impacts of SJFE and GSEE Combinations on MCF7 Cells

In humans, ERα and ERβ are pivotal for regulating various physiological processes. ERα predominantly influences reproductive function, whereas ERβ is involved in the central nervous system, cardiovascular system, and bone health [19]. MCF7 cells are commonly used to assess the estrogenic effects of phytoestrogens, owing to their stable estrogen sensitivity and to the reproducibility of these effects in these cells [20]. Menopausal symptoms arise largely from the decline in estrogen levels, which these receptors mediate. Here, we examined the potential of combinations of these extracts, as estrogen substitutes, to alleviate menopausal symptoms by targeting Erα in particular. The expression of the estrogen-responsive gene pS2 was significantly elevated in MCF7 cells treated with a combination of SJFE and GSEE at a 2:1 ratio (Figure 2A). Coincidently, this combination significantly increased ERα expression and Akt activation (Figure 2B–D). These results suggest the efficacy of this combination in mimicking the effects of estrogen, consistent with previous findings highlighting its estrogen-like activity [21,22]. Here, following the guidelines recommended by the Korea Food and Drug Administration, we focused on the expression of the pS2 gene in MCF7 cells, which are known for their responsiveness to estradiol, as well as on ERα activation and Akt phosphorylation. Although other biomarkers could have been analyzed, the selected markers effectively represent the key aspects of estrogenic activity relevant to our research.

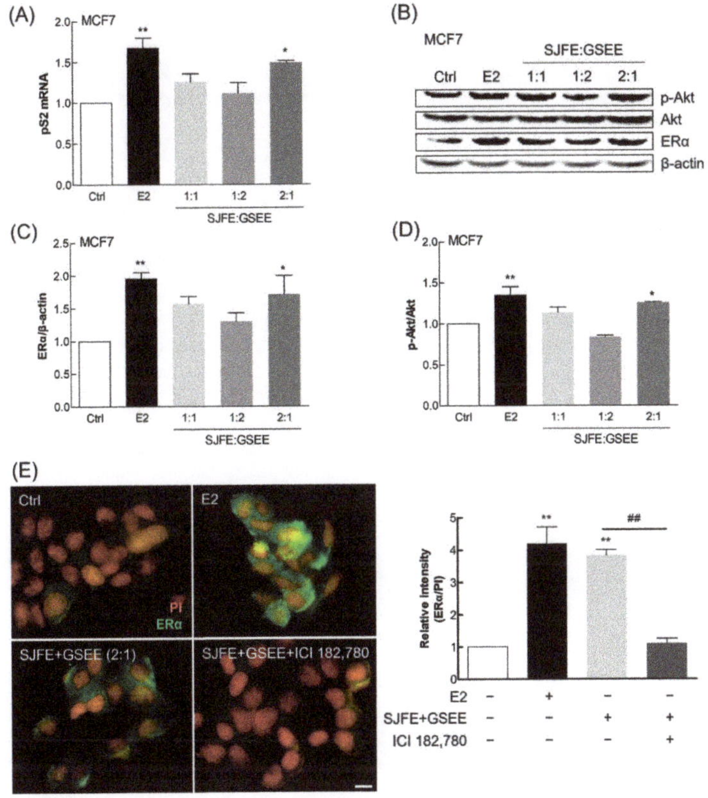

Figure 2. Evaluation of the estrogen-like activity of *Styphnolobium japonicum* fruit extract (SJFE) and germinated soybean embryo extract (GSEE) combinations in MCF7 cells. (**A**) Expression of pS2 mRNA, quantified using qPCR. (**B–D**) Representative immunoblots of ERα, phosphorylated Akt (p-Akt), total Akt, and β-actin. (**E**) Immunocytochemical analysis of ERα expression (left panel). The relative ERα/PI ratio is presented as fold-change relative to the control (right panel). 17β-estradiol (E2) was used as a positive control, and ICI 182,780 was used as an estrogen antagonist. Scale bar, 100 μm. Results from three independent experiments are shown as means ± standard deviation. * $p < 0.05$, ** $p < 0.01$, vs. the control group; ## $p < 0.01$, vs. the indicated group. ER, estrogen receptor; PI, propidium iodide.

Immunocytochemical staining for ERα in MCF7 cells revealed that, even in the absence of estrogen, the SJFE–GSEE complex activated this receptor, indicating the combination's estrogen-like properties (Figure 2E). This activation was inhibited by ICI 182,780, an estrogen antagonist, confirming the competitive specificity of the interaction between the test substances and ERα [23].

These findings demonstrate that the complex extract (SJFE:GSEE, 2:1) can potentially compensate for estrogen deficiency during menopause, enhancing cellular responses in a similar way to estrogen. The complex extract is therefore a promising agent for addressing menopausal symptoms related to estrogen decline, consistent with previous findings [24,25]. These findings underscore the potential application of this extract in treating menopausal symptoms, opening new avenues for natural health-product development and research into women's health.

3.3. Composition of Active Compounds in the SJF–GSE Extract

We examined the extracts' active components and marker compounds, sophoricoside and soyasaponin 1. Sophoricoside exhibits a multifaceted pharmacological profile, including estrogenic, anti-inflammatory, antioxidative, antidiabetic, and immunomodulatory activity [12,26]. Soyasaponins are distinguished by their diverse chemical structures and broad-spectrum health benefits, with soyasaponin 1 noted for its anti-inflammatory, anticarcinogenic, hepatoprotective, and antioxidative properties [10,11,27]. We used HPLC to quantify the levels of these compounds (Figure S5). This revealed substantial levels of sophoricoside and soyasaponin 1 within the extract, at 59.25 ± 0.68 mg/g and 9.09 ± 2.84 mg/g, respectively (Figure 3). These results fall within the range (48–72 mg/g sophoricoside and 8–12 mg/g soyasaponin 1) established when analyzing the combinations of SJFE and GSEE, indicating that the composition of the active compounds is equivalent (Table S3). These high concentrations highlight the extracts' potential roles in ameliorating menopausal symptoms and emphasize the therapeutic relevance of the composition of the complex extract.

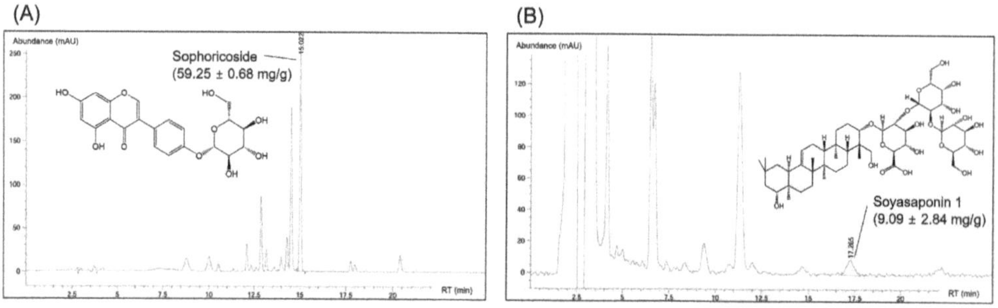

Figure 3. Analysis of active compounds in the *Styphnolobium japonicum* fruit (SJF) and germinated soybean embryo (GSE) complex extract, prepared by blending the SJF and GSE. HPLC was used to analyze sophoricoside (**A**) and soyasaponin 1 (**B**) levels. The compounds were identified and measured by comparing their retention times (RTs) against those of the standards.

3.4. Body Weight and Vasomotor Symptom Improvement

We next evaluated the potential of the complex extract to alleviate menopausal symptoms in the OVX rat model. During the 12-week experimental period, changes in body weight were monitored and vasomotor symptoms were examined as a key indicator of facial flushing during menopause. The estrogen-supplemented positive control group showed the least variation in weight, illustrating the significant role of estrogen in weight regulation after ovariectomy. Notably, the complex extract-treated groups exhibited significantly lower weight gain than the OVX control group (Figure 4A). This is consistent with post-ovariectomy weight-management trends in rodent models, suggesting the effectiveness of the SJF–GSE extract in simulating the weight-regulatory role of estrogen [28,29]. Additionally, we measured changes in vasomotor symptom based on the initial temperature readings (Figure 4B). The rectal temperature of OVX rats changed substantially and remained constant for >2 h. Facial flushing, resulting from capillary congestion in the epidermis, leads to increased skin blood flow, elevated temperatures, and reduced skin moisture [30]. These symptoms are often linked to substantial declines in estrogen levels during menopause, which are believed to reduce the hypothalamic setpoint for body temperature regulation [31]. This suggests that even minor increases in core body temperature could trigger vasomotor symptoms, making their observation a crucial marker for assessing the severity of menopausal symptoms resulting from hormonal depletion. The OVX group exhibited substantial fluctuations in temperature, indicating an unstable post-exercise response. Although the complex extract-treated rats also exhibited temperature variation after exercise, their temperature gradually normalized, following a pattern similar to that observed in the positive control group. This suggests a link between the induced temperature changes in our model and common menopausal symptoms in humans, such as facial flushing and increased body temperature; further, these symptoms were affected by the complex extract, regardless of the dose [32].

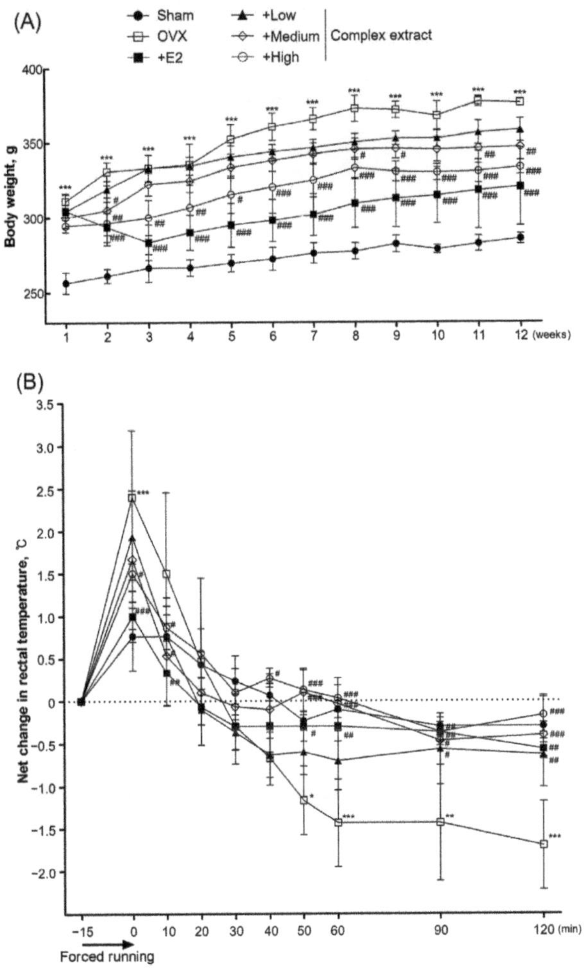

Figure 4. Body weight and vasomotor symptoms in ovariectomized (OVX) rats treated with the *Styphnolobium japonicum* fruit (SJF) and germinated soybean embryo (GSE) complex extract. (**A**) Body weight over 12 weeks. (**B**) Rectal temperature changes during the 120 min post-exercise. The complex extract was prepared by blending SJF and GSE. 17β-estradiol (E2) was used as a positive control. The results are presented as the means ± standard deviation ($n = 7$). * $p < 0.05$, ** $p < 0.01$, *** $p < 0.001$ vs. the sham group; # $p < 0.05$, ## $p < 0.01$, ### $p < 0.001$, vs. the OVX group.

3.5. Postmenopausal Vaginal Symptoms (Vaginal Keratinization)

We examined changes in vaginal epithelial cells, focusing on vaginal cornification as a key indicator. Urogenital symptoms, including vaginal dryness, are prevalent in >50% of menopausal women, primarily owing to hormonal fluctuations during menopause [33]. These hormonal changes often lead to the thinning and cornification of the vaginal lining. To examine the efficacy of our experimental substances in mitigating these symptoms, we analyzed the vaginal epithelial cells from each group. Using Giemsa staining, we quantified the nucleated and cornified cells from vaginal epithelial samples collected using sterile swabs immediately before euthanasia. The proportion of cornified cells serves as an indicator of vaginal health; a higher proportion indicates reduced cornification and improved mucosal flexibility, factors that are essential for maintaining vaginal tissue health during menopause [34,35]. The Medium- and High-dose extract-treated groups displayed higher levels of cornified epithelial cells than the positive control group, indicating amelioration of the vaginal mucosal condition (Figure 5). This suggests that the extracts potentially enhance the flexibility and overall health of the vaginal mucosa in a menopausal model, mirroring the therapeutic benefits observed in other studies [36].

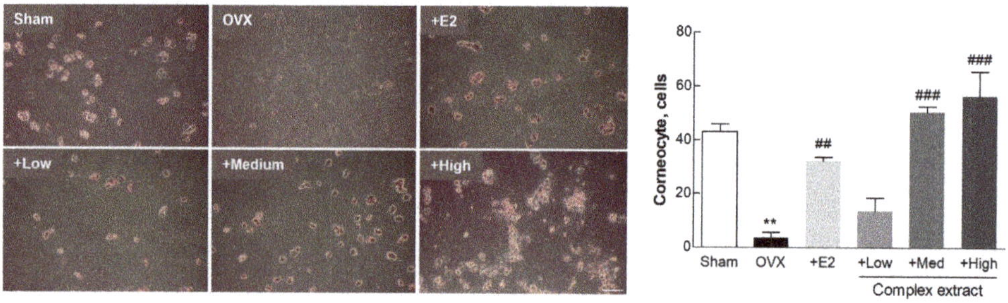

Figure 5. Vaginal cornification in ovariectomized (OVX) rats treated with the *Styphnolobium japonicum* fruit (SJF) and germinated soybean embryo (GSE) complex extract. Representative images of stained vaginal epithelial cells, illustrating the proportion of nucleated and cornified cells (left panel). Counting and comparison of the number of cornified cells (right panel). Scale bar, 100 μm. The complex extract was prepared by blending the SJF and GSE. 17β-estradiol (E2) was used as a positive control. The results are presented as means ± standard deviation ($n = 7$). ** $p < 0.01$, vs. the sham group; ## $p < 0.01$, ### $p < 0.001$, vs. the OVX group.

3.6. Postmenopausal Changes in Uterine and Adipose Tissue

Our investigation into the therapeutic effects of the complex extract on postmenopausal symptoms yielded important insights into changes in uterine and adipose tissue. The complex extract exhibited a favorable safety profile, with no significant changes in organ weight, suggesting its suitability for use without the risk of adverse organ-related effects (Figure S6). Notably, we observed a dose-dependent increase in uterine weight among the treatment groups (Figure 6A), indicating the compensatory estrogenic effects of the complex extract. This increase in uterine weight suggests the potential of this extract to maintain uterine health; this potential is especially relevant under conditions simulating postmenopausal estrogen deficiency [37]. The complex extract-treated groups exhibited an increase in uterine wall thickness (Figure 6B), consistent with our findings on uterine weight, and further supporting the notion that treatment can effectively counteract the negative effects of estrogen deficiency [38]. These findings highlight the therapeutic potential of this extract in mimicking the beneficial effects of estrogen on uterine tissue, thus offering a promising avenue for the management of postmenopausal symptoms.

Following ovariectomy and subsequent treatment, adipose tissue weight was significantly higher in the OVX group than that in the sham group (Figure 7A). This supports the notion that estrogen depletion contributes to increased adiposity. The treatment groups, and especially the High-dose extract-treated group, exhibited adiposity levels comparable to those in the sham group, indicating the complex extract's effectiveness in countering menopause-associated adipocyte proliferation and lipid accumulation. The observed regulation of lipid-droplet size in the complex extract-treated groups (Figure 7B) supports prior findings regarding the role of estrogen in managing body weight and visceral fat after ovariectomy [39]. These results provide substantial evidence of the ability of the SJF–GSE extract to alleviate menopausal symptoms, including increased adiposity, associated with estrogen depletion.

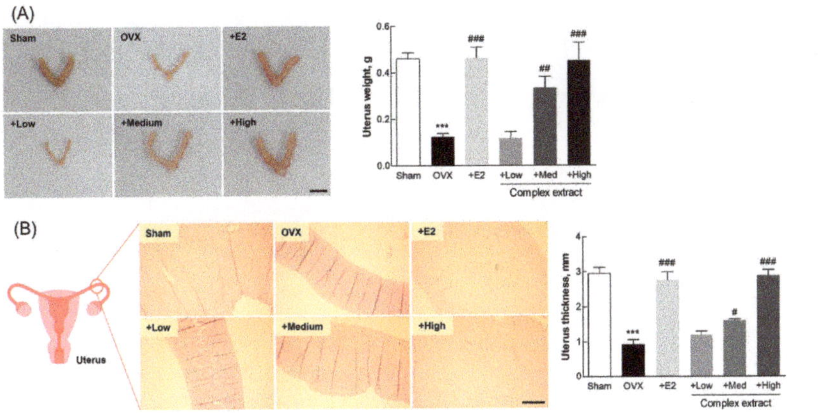

Figure 6. Uterine changes in ovariectomized (OVX) rats treated with the *Styphnolobium japonicum* fruit (SJF) and germinated soybean embryo (GSE) complex extract. (**A**) Representative images of dissected uterine tissue from the rats (left panel). Uterine weight (right panel). Scale bar, 1 cm. (**B**) Representative images of hematoxylin and eosin-stained uterine walls (left panel). Uterine thickness (right panel). Scale bar, 500 μm. The complex extract was prepared by blending the SJF and GSE. 17β-estradiol (E2) was used as a positive control. The results are presented as means ± standard deviation (n = 3). *** $p < 0.001$, vs. the sham group; # $p < 0.05$, ## $p < 0.01$, ### $p < 0.001$, vs. the OVX group.

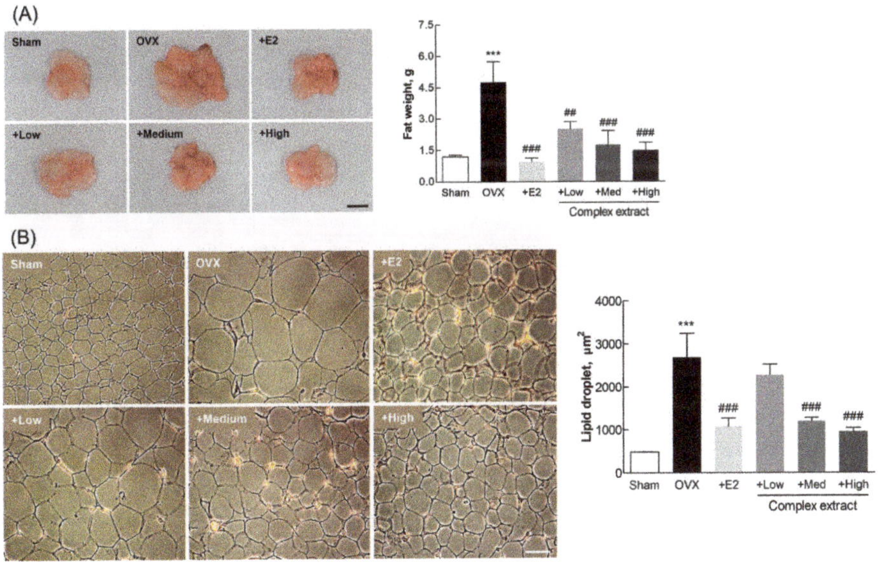

Figure 7. Adipose tissue analysis in ovariectomized (OVX) rats treated with the *Styphnolobium japonicum* fruit (SJF) and germinated soybean embryo (GSE) complex extract. (**A**) Representative images of the perirenal adipose tissue (left panel). Fat weight (right panel). Scale bar, 1 cm. (**B**) Representative images of lipid droplets (left panel). Lipid droplet size (right panel). Scale bar, 100 μm. The complex extract was prepared by blending the SJF and GSE. 17β-estradiol (E2) was used as a positive control. The results are presented as means ± standard deviation (n = 3). *** $p < 0.001$, vs. the sham group; ## $p < 0.01$, ### $p < 0.001$, vs. the OVX group.

These findings reveal the potential of the SJF–GSE complex in mitigating the key symptoms of menopause in OVX rats, thus elucidating its potential for managing postmenopausal conditions in humans. The notable improvements observed in uterine and adipose tissue mark a significant step toward developing effective therapeutic strategies to manage postmenopausal symptoms caused by estrogen deficiency.

3.7. Blood Biochemical Markers

We examined blood biochemical markers in the OVX rats following the 12-week treatment regimen with varying concentrations of the complex extract (Table 2). E2 levels were significantly elevated in the positive control and High-dose extract-treated groups, indicating the estrogen-mimicking potential of these treatments [40]. Levels of ALT, an indicator of liver health, initially increased in the extract-treated groups but declined with increasing dosage, suggesting a dose-dependent beneficial effect [41]. TG levels were reduced in both the E2-and complex extract-treated groups. The LDL levels observed suggest the possible long-term benefits of these treatments on lipid metabolism and cardiovascular health [42]. The LDH levels, which are indicative of tissue damage and serve as a predictive marker for various cancers—especially breast cancer in postmenopausal women—varied significantly with extract concentration. This suggests that ongoing administration of this extract may improve postmenopausal health [43].

Table 2. Blood biochemical indices in ovariectomized (OVX) rats treated with the *Styphnolobium japonicum* fruit (SJF) and germinated soybean embryo (GSE) complex extract.

Parameter	Sham	OVX	+E2	Complex Extract		
				+Low	+Medium	+High
E2 (pg/mL)	24.68 ± 1.14	14.4 ± 0.63 *	29.72 ± 1.4 ##	19.86 ± 1.5	22.47 ± 2.29	33.59 ± 1.88 ##
CTx (pg/mL)	292.97 ± 33.6	548.31 ± 28.81 **	293.39 ± 21.57 ##	365.98 ± 49.36 #	264.35 ± 12.44 ##	196.32 ± 19.91 ##
OC (pg/mL)	7.44 ± 0.34	13.31 ± 0.51 **	5.11 ± 0.7 ###	9.44 ± 0.57 #	7.57 ± 0.58 ##	6.29 ± 0.17 ###
Ca (mg/dL)	10.45 ± 0.05	10.55 ± 0.55	11.15 ± 1.25	10.2 ± 0.1	10.45 ± 0.05	10.8 ± 0.5
P (mg/dL)	3.65 ± 0.05	2.15 ± 0.15 **	3.4 ± 0.2 #	2.3 ± 0.2	3.35 ± 0.15 #	3.75 ± 0.25 ##
TG (mg/dL)	49 ± 5	139.25 ± 0.75 **	82.9 ± 7.1 #	94.4 ± 13.6	74.05 ± 13.45	56.1 ± 2.1 #
HDL (mg/dL)	48.3 ± 1.8	41.75 ± 5.25	40.03 ± 1.02	36.5 ± 1.5	43.5 ± 1.5	50.4 ± 2.6
LDL (mg/dL)	5.5 ± 0.5	10.5 ± 0.5 **	5.5 ± 0.5 ##	10.5 ± 0.5	10 ± 1	7.5 ± 0.5 #
ALT (U/L)	70.3 ± 2.8	102.05 ± 2.15 *	88.55 ± 6.15	105.95 ± 4.05	87.95 ± 9.25	63.35 ± 2.05 #
LDH (U/L)	318.25 ± 6.25	945.25 ± 80.25 **	684.5 ± 79.5	396.25 ± 20.25 ##	210.5 ± 9.5 ###	227.5 ± 57.5 ###

OVX rats were orally administered the complex extract at 25 mg/kg (low), 50 mg/kg (medium), and 100 mg/kg for 12 weeks. 17β-estradiol (E2) was used as a positive control. * $p < 0.05$, ** $p < 0.01$, vs. the sham group; # $p < 0.05$, ## $p < 0.01$, ### $p < 0.001$, vs. the OVX group. E2, estradiol; CTx, C-terminal telopeptide; OC, osteocalcin; Ca, calcium; P, phosphorus; TG, triglycerides; HDL, high-density lipoprotein; LDL, low-density lipoprotein; ALT, alanine aminotransferase; LDH, lactate dehydrogenase.

The effect of the extracts on the bone health markers osteocalcin and CTx were examined. Osteocalcin and CTx levels were significantly elevated after ovariectomy, but were decreased in the positive control and complex extract-treated groups, exhibiting a dose-dependent response. When comparing the serum levels of minerals associated with bone health, Ca levels were not significantly altered; P levels, in contrast, were decreased in the OVX group and were dose-dependently elevated in the extract-treated groups, with the High-concentration and sham groups exhibiting similar P levels. This suggests that the complex extract potentially counteracts the bone loss associated with estrogen depletion [44,45]. These results strongly suggest that this complex extract can effectively modulate key biochemical markers in post-OVX rat models, thus highlighting their potential for managing the symptoms and health concerns related to menopause.

3.8. Evaluation of Femoral Bone Features: Gene Expression, Histological Staining, and Imaging Analysis

We investigated the effects of the SJF–GSE extract on bone features in the OVX rats, focusing on (1) the expression of genes related to bone resorption and formation, (2) histological changes, and (3) and bone mineralization (Figure 8). The decline in estrogen levels following menopause boosts the formation of osteoclasts, bone-resorbing cells involved in bone remodeling [46]. We observed significant downregulation in the expression of RANK and interleukin-1β (IL-1β), markers associated with osteoclast activity [47], in all of the extract-treated groups, suggesting a reduction in bone resorption (Figure 8A,B). The expression of tartrate-resistant acid phosphatase (TRAP), which is indicative of osteoclast differentiation [48,49], was downregulated in the extract-treated groups, indicating the potential of the complex extract to promote bone health by reducing osteoclastogenesis (Figure 8C). Histological assessment via H&E staining revealed significant preservation of bone density by the extract complex, indicating its efficacy in mitigating osteoporotic changes and estrogen deficits (Figure 8D).

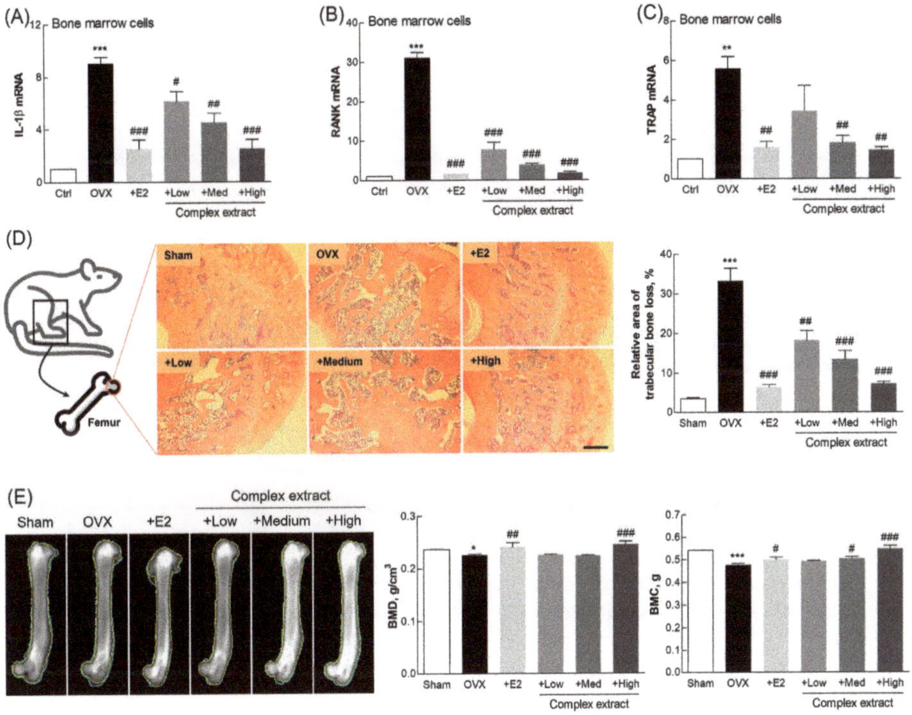

Figure 8. Bone features and gene expression profiles in ovariectomized (OVX) rats treated with the *Styphnolobium japonicum* fruit (SJF) and germinated soybean embryo (GSE) complex extract. (**A–C**) Expression of bone resorption-related genes, evaluated using qPCR, in femoral bone-derived bone marrow cells. (**D**) Hematoxylin and eosin-stained transverse sections of femur (left panel). Trabecular bone loss area (right panel). (**E**) Representative X-ray images of femoral bone (left panel). Bone mineral density (BMD) and bone mineral content (BMC) (right panel). Scale bar, 500 μm. The complex extract was prepared by blending the SJF and GSE. 17β-estradiol (E2) was used as a positive control. The results are presented as means ± standard deviation ($n = 3$). * $p < 0.05$, ** $p < 0.01$, *** $p < 0.001$, vs. the sham group; # $p < 0.05$, ## $p < 0.01$, ### $p < 0.001$, vs. the OVX group. IL-1β, interleukin 1β; RANK, receptor activator of nuclear factor κB; TRAP, tartrate-resistant acid phosphatase.

BMC and BMD were examined using X-ray imaging to visualize changes in bone structure. This revealed substantially greater bone density at higher extract concentrations (Figure 8E). The complex extract-treated group (and especially the High-dose group) exhibited higher BMC and BMD than the OVX group, indicating improved bone strength and reduced fracture risk. These findings, consistent with prior findings, support the osteoprotective effects of phytoestrogen [25]. This further confirms the phytoestrogen-like action of the complex extract [50,51].

These findings reveal the benefits and potential therapeutic value of the SJF–GSE complex extract on bone health and in the alleviation of estrogen depletion-associated postmenopausal symptoms. This product, which substantially downregulated osteoclastogenesis-related gene expression and improved the histological structure and mineralization profiles represents an important advancement in developing efficacious treatments for postmenopausal conditions.

Notably, it is important to acknowledge that the biological effects observed in our study are likely influenced by the inherent complexity of the extracts. These extracts are not simple one-component systems but rather complex mixtures of various biomolecules, each contributing to the overall activity. Interactions among these components can lead to synergistic or additive effects, enhancing the therapeutic potential of the extracts [52]. Conversely, antagonistic interactions may also occur, where certain components could suppress the activity of others, impacting the overall efficacy [53]. Moreover, the biological effects of these extracts are closely tied to the extraction methods used, as these methods determine the final composition and activity of the extracts. Although our study primarily focused on the overall effects of the extracts, future research should aim to dissect these interactions more thoroughly to better understand how they contribute to the observed outcomes.

4. Conclusions

The complex extract of *S. japonicum* and germinated soybean embryo demonstrated significant therapeutic efficacy in alleviating postmenopausal symptoms, including improved bone density, reduced vasomotor symptoms, and enhanced vaginal health. The extract's estrogen-like effects and antioxidant properties highlight its potential as a natural alternative to hormone replacement therapy. These findings support the development of plant-based therapies for managing menopausal symptoms.

Supplementary Materials: The following supporting items can be downloaded at: https://www.mdpi.com/article/10.3390/nu16193297/s1, Figure S1: Effects of SJFE, GSEE, and their combinations on cell viability; Figure S2: Anti-oxidative effects of SJFE, GSEE, and their combinations, evaluated based on their DPPH radical scavenging activity; Figure S3: Effects of SJFE, GSEE, and their combinations on nitric oxide (NO) production in RAW 264.7 cells; Figure S4: Effect of SJFE, GSEE, and their combinations on RANKL gene expression in MG63 cells; Figure S5: HPLC analysis of the sophoricoside and soyasaponin 1 standards; Figure S6: Organ safety assessment in ovariectomized (OVX) rats treated with the complex extract; Table S1: Primer sequences for gene expression analysis; Table S2: Antibody list for western blot analysis; Table S3: Quantitative analysis of the active compounds in the SJFE and GSEE mixtures and in the final complex extract.

Author Contributions: Conceptualization, J.-C.J. and S.-S.J.; methodology, H.-H.S. and K.-M.K.; software, K.D. and J.-Y.P.; validation, K.-M.K.; formal analysis, J.-W.A. and S.-K.J.; investigation, H.-S.K. and J.-Y.P.; resources, J.-C.J.; data curation, J.-W.A., H.-S.K. and K.D.; writing—original draft preparation, J.-W.A. and H.-S.K.; writing—review and editing, Y.-M.Y., J.-C.J. and S.-S.J.; visualization, J.-W.A. and H.-H.S.; supervision, S.-S.J.; project administration, S.-K.J. and Y.-M.Y.; funding acquisition, J.-C.J. All authors have read and agreed to the published version of the manuscript.

Funding: This research was financially supported by the Ministry of Trade, Industry and Energy, Republic of Korea, under the "World Class Plus Program (R&D, P0017150)" and "Middle Market Enterprises Global Research R&D Project (R&D, P0000784)", supervised by the Korea Institute for Advancement of Technology (KIAT).

Institutional Review Board Statement: All animal experiments were approved by the Institutional Animal Care and Use Committee of Gangneung-Wonju National University, Gangneung, Republic of Korea (GWNU-2020-18, approved on 19 May 2020), and were conducted in accordance with the relevant guidelines and regulations.

Informed Consent Statement: Not applicable.

Data Availability Statement: The original contributions presented in the study are included in the article/Supplementary Material, further inquiries can be directed to the corresponding author.

Conflicts of Interest: Authors Hee-Hyun Shin, Kyung-Mi Kim and Jae-Chul Jung were employed by the company Life Science Research Institute, NOVAREX Co., Ltd. The remaining authors declare that the research was conducted in the absence of any commercial or financial relationships that could be construed as a potential conflict of interest.

References

1. Duralde, E.R.; Sobel, T.H.; Manson, J.E. Management of perimenopausal and menopausal symptoms. *Brit Med. J.* **2023**, *382*, e072612. [CrossRef] [PubMed]
2. Fait, T. Menopause hormone therapy: Latest developments and clinical practice. *Drugs Context* **2019**, *8*, 212551. [CrossRef]
3. Chen, Z.; Qian, F.; Hu, Y.; Voortman, T.; Li, Y.; Rimm, E.B.; Sun, Q. Dietary phytoestrogens and total and cause-specific mortality: Results from 2 prospective cohort studies. *Am. J. Clin. Nutr.* **2023**, *117*, 130–140. [CrossRef]
4. Rowe, I.J.; Baber, R.J. The effects of phytoestrogens on postmenopausal health. *Climacteric* **2021**, *24*, 57–63. [CrossRef] [PubMed]
5. Kim, M.R.; Kim, H.J.; Yu, S.H.; Lee, B.S.; Jeon, S.Y.; Lee, J.J.; Lee, Y.C. Combination of red clover and hops extract improved menopause symptoms in an ovariectomized rat model. *Evid. Based Complement. Altern. Med.* **2020**, *2020*, 7941391. [CrossRef] [PubMed]
6. Mahdavian, M.; Mirzaii Najmabadi, K.; Hosseinzadeh, H.; Mirzaeian, S.; Badiee Aval, S.; Esmaeeli, H. Effect of the mixed herbal medicines extract (fennel, chamomile, and saffron) on menopause syndrome: A randomized controlled clinical trial. *J. Caring Sci.* **2019**, *8*, 181–189. [CrossRef] [PubMed]
7. Bau, H.M.; Villaume, C.; Méjean, L. Effects of soybean (*Glycine max*) germination on biologically active components, nutritional values of seeds, and biological characteristics in rats. *Nahrung* **2000**, *44*, 2–6. [CrossRef]
8. Weaver, C.M.; Martin, B.R.; Jackson, G.S.; McCabe, G.P.; Nolan, J.R.; McCabe, L.D.; Barnes, S.; Reinwald, S.; Boris, M.E.; Peacock, M. Antiresorptive effects of phytoestrogen supplements compared with estradiol or risedronate in postmenopausal women using (41)Ca methodology. *J. Clin. Endocrinol. Metab.* **2009**, *94*, 3798–3805. [CrossRef]
9. Jeong, H.; Lee, J.; Kim, S.; Park, S.; Yang, H.; Ahn, B.H.; Jang, C.Y.; Jeong, H.C.; Lee, S.J.; Kim, S.L.; et al. Characterization of Soybean Germinated Embryo Extract as an Estrogen Receptor Subtype-Selective and Tissue-Specific Modulator. *J. Med. Food* **2019**, *22*, 186–195. [CrossRef]
10. Lee, K.-S.; Woo, S.-Y.; Lee, M.-J.; Kim, H.Y.; Ham, H.; Lee, D.-J.; Choi, S.-W.; Seo, W.D. Isoflavones and soyasaponins in the germ of Korean soybean [*Glycine max* (L.) Merr.] cultivars and their compound-enhanced BMP-2-induced bone formation. *Appl. Biol. Chem.* **2020**, *63*, 26. [CrossRef]
11. Kamo, S.; Suzuki, S.; Sato, T. Comparison of bioavailability (I) between soyasaponins and soyasapogenols, and (II) between group A and B soyasaponins. *Nutrition* **2014**, *30*, 596–601. [CrossRef] [PubMed]
12. Shi, P.; Liao, J.; Duan, T.; Wu, Q.; Huang, X.; Pei, X.; Wang, C. Chemical composition and pharmacological properties of *Flos sophorae immaturus*, *Flos sophorae* and *Fructus sophorae*: A review. *J. Future Foods* **2023**, *3*, 330–339. [CrossRef]
13. Ahn, J.W.; Jang, S.K.; Jo, B.R.; Kim, H.S.; Park, J.Y.; Park, H.Y.; Yoo, Y.M.; Joo, S.S. A therapeutic intervention for Alzheimer's disease using ginsenoside Rg3: Its role in M2 microglial activation and non-amyloidogenesis. *J. Physiol. Pharmacol.* **2021**, *72*, 185–193. [CrossRef]
14. Stout Steele, M.; Bennett, R.A. Clinical technique: Dorsal ovariectomy in rodents. *J. Exot. Pet. Med.* **2011**, *20*, 222–226. [CrossRef]
15. Muschter, D.; Göttl, C.; Vogel, M.; Grifka, J.; Straub, R.H.; Grässel, S. Reactivity of rat bone marrow-derived macrophages to neurotransmitter stimulation in the context of collagen II-induced arthritis. *Arthritis Res. Ther.* **2015**, *17*, 169. [CrossRef]
16. Leanza, G.; Conte, C.; Cannata, F.; Isgrò, C.; Piccoli, A.; Strollo, R.; Quattrocchi, C.C.; Papalia, R.; Denaro, V.; Maccarrone, M.; et al. Oxidative stress in postmenopausal women with or without obesity. *Cells* **2023**, *12*, 1137. [CrossRef]
17. Doshi, S.B.; Agarwal, A. The role of oxidative stress in menopause. *J. Midlife Health* **2013**, *4*, 140–146. [CrossRef]
18. Mohamad, N.V.; Ima-Nirwana, S.; Chin, K.Y. Are oxidative stress and inflammation mediators of bone loss due to estrogen deficiency? A review of current evidence. *Endocr. Metab. Immune Disord. Drug Targets* **2020**, *20*, 1478–1487. [CrossRef] [PubMed]
19. Paterni, I.; Granchi, C.; Katzenellenbogen, J.A.; Minutolo, F. Estrogen receptors alpha (ERα) and beta (ERβ): Subtype-selective ligands and clinical potential. *Steroids* **2014**, *90*, 13–29. [CrossRef] [PubMed]
20. Lee, D.; Kim, Y.M.; Chin, Y.W.; Kang, K.S. Schisandrol A exhibits estrogenic activity via estrogen receptor α-dependent signaling pathway in estrogen receptor-positive breast cancer cells. *Pharmaceutics* **2021**, *13*, 1082. [CrossRef]
21. Moriarty, K.; Kim, K.H.; Bender, J.R. Estrogen receptor-mediated rapid signaling. *Endocrinology* **2006**, *147*, 5557–5563. [CrossRef]

22. Echeverria, V.; Echeverria, F.; Barreto, G.E.; Echeverría, J.; Mendoza, C. Estrogenic plants: To prevent neurodegeneration and memory loss and other symptoms in women after menopause. *Front. Pharmacol.* **2021**, *12*, 644103. [CrossRef]
23. Movérare-Skrtic, S.; Börjesson, A.E.; Farman, H.H.; Sjögren, K.; Windahl, S.H.; Lagerquist, M.K.; Andersson, A.; Stubelius, A.; Carlsten, H.; Gustafsson, J.; et al. The estrogen receptor antagonist ICI 182,780 can act both as an agonist and an inverse agonist when estrogen receptor α AF-2 is modified. *Proc. Natl. Acad. Sci. USA* **2014**, *111*, 1180–1185. [CrossRef] [PubMed]
24. He, X.; Bai, Y.; Zhao, Z.; Wang, X.; Fang, J.; Huang, L.; Zeng, M.; Zhang, Q.; Zhang, Y.; Zheng, X. Local and traditional uses, phytochemistry, and pharmacology of *Sophora japonica* L.: A review. *J. Ethnopharmacol.* **2016**, *187*, 160–182. [CrossRef]
25. Canivenc-Lavier, M.C.; Bennetau-Pelissero, C. Phytoestrogens and health effects. *Nutrients* **2023**, *15*, 317. [CrossRef] [PubMed]
26. Aly, S.H.; Elbadry, A.M.M.; El-Shazly, M.; Hwang, T.-L. Exploring the potential role of genus *Sophora* in the management of osteoporosis: A phytochemical and biological review. *Front. Nat. Prod.* **2023**, *2*, 1302371. [CrossRef]
27. Guang, C.; Chen, J.; Sang, S.; Cheng, S. Biological functionality of soyasaponins and soyasapogenols. *J. Agric. Food Chem.* **2014**, *62*, 8247–8255. [CrossRef] [PubMed]
28. Wang, Y.; Wang, Y.; Liu, L.; Cui, H. Ovariectomy induces abdominal fat accumulation by improving gonadotropin-releasing hormone secretion in mouse. *Biochem. Biophys. Res. Commun.* **2022**, *588*, 111–117. [CrossRef]
29. Wellberg, E.A.; Corleto, K.A.; Checkley, L.A.; Jindal, S.; Johnson, G.; Higgins, J.A.; Obeid, S.; Anderson, S.M.; Thor, A.D.; Schedin, P.J.; et al. Preventing ovariectomy-induced weight gain decreases tumor burden in rodent models of obesity and postmenopausal breast cancer. *Breast Cancer Res.* **2022**, *24*, 42. [CrossRef]
30. Kobayashi, T.; Tamura, M.; Hayashi, M.; Katsuura, Y.; Tanabe, H.; Ohta, T.; Komoriya, K. Elevation of tail skin temperature in ovariectomized rats in relation to menopausal hot flushes. *Am. J. Physiol. Regul. Integr. Comp. Physiol.* **2000**, *278*, R863–R869. [CrossRef] [PubMed]
31. Zhang, Z.; DiVittorio, J.R.; Joseph, A.M.; Correa, S.M. The effects of estrogens on neural circuits that control temperature. *Endocrinology* **2021**, *162*, bqab087. [CrossRef]
32. Lethaby, A.; Marjoribanks, J.; Kronenberg, F.; Roberts, H.; Eden, J.; Brown, J. Phytoestrogens for menopausal vasomotor symptoms. *Cochrane Database Syst. Rev.* **2013**, *2013*, CD001395. [CrossRef] [PubMed]
33. Angelou, K.; Grigoriadis, T.; Diakosavvas, M.; Zacharakis, D.; Athanasiou, S. The genitourinary syndrome of menopause: An overview of the recent data. *Cureus* **2020**, *12*, e7586. [CrossRef] [PubMed]
34. Lee, S.; Jung, D.H.; Park, M.; Yeon, S.W.; Jung, S.H.; Yun, S.I.; Park, H.O.; Yoo, W. The effect of *Lactobacillus gasseri* BNR17 on postmenopausal symptoms in ovariectomized rats. *J. Microbiol. Biotechnol.* **2021**, *31*, 1281–1287. [CrossRef]
35. Parhizkar, S.; Latiff, L.A. Supplementary health benefits of linoleic acid by improvement of vaginal cornification of ovariectomized rats. *Adv. Pharm. Bull.* **2013**, *3*, 31–36. [CrossRef] [PubMed]
36. Anderson, D.J.; Marathe, J.; Pudney, J. The structure of the human vaginal stratum corneum and its role in immune defense. *Am. J. Reprod. Immunol.* **2014**, *71*, 618–623. [CrossRef] [PubMed]
37. Hebbar, S.; Chaya, V.; Rai, L.; Ramachandran, A. Factors influencing endometrial thickness in postmenopausal women. *Ann. Med. Health Sci. Res.* **2014**, *4*, 608–614. [CrossRef]
38. Yu, K.; Huang, Z.Y.; Xu, X.L.; Li, J.; Fu, X.W.; Deng, S.L. Estrogen receptor function: Impact on the human endometrium. *Front. Endocrinol.* **2022**, *13*, 827724. [CrossRef]
39. Babaei, P.; Mehdizadeh, R.; Ansar, M.M.; Damirchi, A. Effects of ovariectomy and estrogen replacement therapy on visceral adipose tissue and serum adiponectin levels in rats. *Menopause Int.* **2010**, *16*, 100–104. [CrossRef]
40. Domínguez-López, I.; Yago-Aragón, M.; Salas-Huetos, A.; Tresserra-Rimbau, A.; Hurtado-Barroso, S. Effects of dietary phytoestrogens on hormones throughout a human lifespan: A review. *Nutrients* **2020**, *12*, 2456. [CrossRef]
41. Shim, J.J.; Kim, J.W.; Oh, C.H.; Lee, Y.R.; Lee, J.S.; Park, S.Y.; Kim, B.H.; Oh, I.H. Serum alanine aminotransferase level and liver-related mortality in patients with chronic hepatitis B: A large national cohort study. *Liver Int.* **2018**, *38*, 1751–1759. [CrossRef] [PubMed]
42. Kosmas, C.E.; Rodriguez Polanco, S.; Bousvarou, M.D.; Papakonstantinou, E.J.; Peña Genao, E.; Guzman, E.; Kostara, C.E. The triglyceride/high-density lipoprotein cholesterol (TG/HDL-C) ratio as a risk marker for metabolic syndrome and cardiovascular disease. *Diagnostics* **2023**, *13*, 929. [CrossRef] [PubMed]
43. Liu, D.; Wang, D.; Wu, C.; Zhang, L.; Mei, Q.; Hu, G.; Long, G.; Sun, W. Prognostic significance of serum lactate dehydrogenase in patients with breast cancer: A meta-analysis. *Cancer Manag. Res.* **2019**, *11*, 3611–3619. [CrossRef] [PubMed]
44. Schini, M.; Vilaca, T.; Gossiel, F.; Salam, S.; Eastell, R. Bone turnover markers: Basic biology to clinical applications. *Endocr. Rev.* **2023**, *44*, 417–473. [CrossRef]
45. Goretti Penido, M.; Alon, U.S. Phosphate homeostasis and its role in bone health. *Pediatr. Nephrol.* **2012**, *27*, 2039–2048. [CrossRef]
46. Møller, A.M.J.; Delaissé, J.M.; Olesen, J.B.; Madsen, J.S.; Canto, L.M.; Bechmann, T.; Rogatto, S.R.; Søe, K. Aging and menopause reprogram osteoclast precursors for aggressive bone resorption. *Bone Res.* **2020**, *8*, 27. [CrossRef]
47. Jin Hee, P.; Na Kyung, L.; Soo Young, L. Current understanding of RANK signaling in osteoclast differentiation and maturation. *Mol. Cells* **2017**, *40*, 706–713. [CrossRef]
48. Hayman, A.R. Tartrate-resistant acid phosphatase (TRAP) and the osteoclast/immune cell dichotomy. *Autoimmunity* **2008**, *41*, 218–223. [CrossRef]
49. Wang, L.T.; Chen, L.R.; Chen, K.H. Hormone-related and drug-induced osteoporosis: A cellular and molecular overview. *Int. J. Mol. Sci.* **2023**, *24*, 5814. [CrossRef]

50. Abdi, F.; Alimoradi, Z.; Haqi, P.; Mahdizad, F. Effects of phytoestrogens on bone mineral density during the menopause transition: A systematic review of randomized, controlled trials. *Climacteric* **2016**, *19*, 535–545. [CrossRef]
51. Sansai, K.; Na Takuathung, M.; Khatsri, R.; Teekachunhatean, S.; Hanprasertpong, N.; Koonrungsesomboon, N. Effects of isoflavone interventions on bone mineral density in postmenopausal women: A systematic review and meta-analysis of randomized controlled trials. *Osteoporos. Int.* **2020**, *31*, 1853–1864. [CrossRef] [PubMed]
52. Wang, X.; Ma, Y.; Xu, Q.; Shikov, A.N.; Pozharitskaya, O.N.; Flisyuk, E.V.; Liu, M.; Li, H.; Vargas-Murga, L.; Duez, P. Flavonoids and saponins: What have we got or missed? *Phytomedicine* **2023**, *109*, 154580. [CrossRef] [PubMed]
53. Caesar, L.K.; Cech, N.B. Synergy and antagonism in natural product extracts: When 1 + 1 does not equal 2. *Nat. Prod. Rep.* **2019**, *36*, 869–888. [CrossRef] [PubMed]

nutrients

MDPI

Article

EZH2-Mediated H3K27 Trimethylation in the Liver of Mice Is an Early Epigenetic Event Induced by High-Fat Diet Exposure

Giulia Pinton [1], Mattia Perucca [1], Valentina Gigliotti [1], Elena Mantovani [1], Nausicaa Clemente [2], Justyna Malecka [1], Gabriela Chrostek [1,†], Giulia Dematteis [1], Dmitry Lim [1], Laura Moro [1,*] and Fausto Chiazza [1,*]

1 Department of Pharmaceutical Sciences, Università del Piemonte Orientale (UPO), Largo Donegani 2, 28100 Novara, Italy; giulia.pinton@uniupo.it (G.P.); valentina.gigliotti@uniupo.it (V.G.); gabriela.chrostek98@gmail.com (G.C.); giulia.dematteis@uniupo.it (G.D.)

2 Department of Health Sciences, Interdisciplinary Research Center of Autoimmune Diseases (IRCAD), Università del Piemonte Orientale (UPO), Via Solaroli 17, 28100 Novara, Italy

* Correspondence: laura.moro@uniupo.it (L.M.); fausto.chiazza@uniupo.it (F.C.)

† Current address: Department of Pharmacological and Biomolecular Sciences, Università degli Studi di Milano, Via Balzaretti 9, 20133 Milan, Italy.

Abstract: Background/Objectives: Methyltransferase EZH2-mediated H3K27me3 is involved in liver inflammation and fibrosis, but its role in hepatic metabolic derangements is not yet clearly defined. We investigated if a high-fat diet (HFD) induced early changes in EZH2 expression and H3K27 me3 in the liver of mice. Methods: Five-week-old mice were fed an HFD or a low-fat diet (Control) for 2 weeks (2 W) or 8 weeks (8 W). Body weight was recorded weekly. Glycemia and oral glucose tolerance were assessed at baseline and after 2 W–8 W. Finally, livers were collected for further analysis. Results: As expected, mice that received 8 W HFD showed an increase in body weight, glycemia, and liver steatosis and an impairment in glucose tolerance; no alterations were observed in 2 W HFD mice. Eight weeks of HFD caused hepatic EZH2 nuclear localization and increased H3 K27me3; surprisingly, the same alterations occurred in 2 W HFD mice livers, even before overweight onset. We demonstrated that selective EZH2 inhibition reduced H3K27me3 and counteracted lipid accumulation in HUH-7 cells upon palmitic acid treatment. Conclusions: In conclusion, we point to EZH2/H3K27me3 as an early epigenetic event occurring in fatty-acid-challenged livers both in vivo and in vitro, thus establishing EZH2 as a potential pharmacological target for metabolic derangements.

Keywords: high-fat diet; H3K27 trimethylation; methyltransferase EZH2

1. Introduction

A high-fat diet (HFD) triggers multiorgan impairment and is associated with detrimental effects on human health. The liver exerts crucial metabolic functions, including lipid metabolism, and is one of the most affected tissues in the context of metabolic derangements. Intake of diet-derived saturated lipids causes insulin resistance, promotes metabolic-dependent inflammation (referred to as metaflammation), leads to fibrosis, alters expression of lipogenic genes, and determines organ weight gain with significant steatosis development [1–3].

Recent studies have increasingly explored the hypothesis that chronic long-term hypercaloric diet supply can affect gene expression programs through epigenetic mechanisms and examined their involvement in the progression of metabolic disorders. Fat accumulation, inflammation, oxidative stress, and fibrosis can indeed be regulated by epigenetic mechanisms, including histone post-translational modifications, DNA methylation, and non-coding RNAs [4,5].

Methyltransferase EZH2 (Enhancer of Zeste Homolog 2), the catalytic subunit of the Polycomb Repressor Complex 2 (PCR2), regulates gene expression via trimethylation of lysine 27 on histone H3 (H3K27me3) [6]. The main function of EZH2 is to suppress gene expression and induce chromatin compaction [7]. Notably, EZH2 overexpression has been mainly described in fibrosis and cancer, showing a positive correlation with their progression [6,8]. Recently, the role of EZH2 besides in cancer has been highlighted in different organs, including the liver [8]. Lee and colleagues demonstrated that EZH2 inhibition reduces liver inflammation and fibrosis in preclinical models of advanced non-alcoholic steatohepatitis [9], resulting in decreased expression of IL-6 (Interleukin-6), IL-1β, Interferon-γ (INF-γ), transforming growth factor β (TGF- β), and connective tissue growth factor (CTGF). However, to date, the epigenetic alterations that occur in the liver in response to HFD have been poorly investigated.

Our study aims to determine whether the EZH2/H3K27me3 axis could represent a very early epigenetic marker of liver impairment upon HFD feeding as an early detection of fat accumulation in the liver. Early intervention strategies could be important to prevent fatty liver damage and to improve hepatic-related outcome. Specifically, we investigate the correlation between the onset of metabolic derangements and H3K27me3 modulation in a preclinical model of diet-induced dysmetabolism (8 weeks of HFD feeding), and subsequently, we try to understand how early this correlation occurs (2 weeks of HFD feeding).

2. Materials and Methods

2.1. Animal Model

Adult male C57BL/6 J mice of four weeks of age were utilized. Mice, kept at 3–5/cage with access to water and food ad libitum, were housed in a light-controlled (12 h light, 12 h dark) and temperature-controlled (22–24 °C) room in high-efficiency particulate air (HEPA)-filtered Thoren units (Thoren Caging Systems) at the University of Piemonte Orientale animal facility, Novara, Italy. Animal care and handling were performed in accordance with the Italian law on animal care (D.L. 26/2014) as well as the European Directive (2010/63/UE) and approved by the Organismo Preposto al Benessere Animale (OPBA) of University of Piemonte Orientale, Novara, Italy (DB064.61).

2.1.1. Diet Administration

A timeline representation of the experimental design can be found in Figure S1A. Thirty-eight animals started receiving an LFD (low-fat diet, 13% kcal from fat, 67% kcal from carbohydrates, 20% kcal from proteins, (Laboratori Piccioni, Gessate, Milan, Italy) instead of the normal chow diet provided by the animal facility from the fourth week of age to get used to a refined diet [10]. After one week (five weeks of age), animals were randomly divided into 4 groups: two groups continued to be fed a low-fat diet (control groups) for eight (n = 9) or two (n = 9) weeks, while the other two groups of animals were fed with an HFD (60% kcal from fat, 21% kcal from carbohydrates, 19% kcal from proteins, Laboratori Piccioni) (HFD groups) for eight (n = 10) or two (n = 10) weeks (Figure S1B). Animal body weight, food, water, and caloric intake were recorded weekly. At the end of the diet regiments, mice were euthanized, and livers were collected and preserved in an optimal cutting temperature (OCT) compound or snap frozen at −80 °C for morphological and bio-molecular analyses.

2.1.2. Oral Glucose Tolerance Test

The oral glucose tolerance test (OGTT) was performed before the beginning of the experiment and after 2 and 8 weeks of diet supplementation. Glucose (2 g/kg) was administered by oral gavage after a fasting period of 18 h. The concentrations of serum glucose were measured with a conventional glucometer (GlucoMen LX kit, Menarini Diagnostics, Bagno a Ripoli, Florence, Italy) before glucose administration and after 15, 30, 60, and 120 min.

2.1.3. Oil Red O Tissue Staining and Analysis

OCT (optimal cutting temperature)-embedded livers were cryo-sectioned at −26 °C (thickness of 8 μm) using Superfrost Slide (Epredia, Breda, Netherlands) and rinsed in PBS.

Oil Red O solution (0.5% in isopropanol, Sigma-Aldrich, Darmstadt, Germany) was used to determine lipid droplets in the liver to a final 3:2 concentration in deionized water. Sections were stained for 15 min and then rinsed in water. Slides were scanned by ZEISS Axioscan 7 (Carl Zeiss Microscopy, Jena, Germany) (setting: REF_BF H&E Slide 40x program). Quantitative analyses were executed using Qu-Path 0.5.0. and ImageJ software 1.8.0_345. At least 50 areas per sample were analyzed. The setting used for analysis was: HUE 0–255; saturation 145–255; brightness 145–255. The stained area was calculated as a percentage of the total liver area analyzed.

2.2. Reagents and Antibodies

The polyclonal antibodies specific to EZH2 (at a 1:1000 dilution), histone H3 trimethyl-lysine 27 (H3K27me3, 1:1000), histone H3 (1:2000), and the monoclonal antibody specific to KDM6B (1:1000) were purchased from Active Motif (La Hulpe, Belgium). Anti-rabbit IgG and anti-mouse IgG peroxidase-conjugated antibodies and reagents, including palmitic acid, were from Sigma-Aldrich (St. Louis, MO, USA). The nitrocellulose membrane and ECL were bought from Bio-Rad (Hercules, CA, USA). The EZH2-selective inhibitor, EPZ-6438, was from Selleckchem (Houston, TX, USA).

2.3. Cell Cultures

HUH-7 cells, derived from human hepatocellular carcinoma (provided by CLS, Cell Lines Service GmbH, Wageningen, Netherlands), were maintained in DMEM (Sigma-Aldrich, Cat. No. D5671) supplemented with 10% fetal bovine serum (Gibco, ThermoFisher, Waltham, Massachusetts, USA) Cat. No. 10270), 2 mM L-glutamine (Sigma-Aldrich), and 1% penicillin/streptomycin solution (Sigma-Aldrich)). When 80% confluence was reached, cells were plated for experiments. At 24 h after plating, cells were treated with palmitic acid (PA, 100 μM, PA group), PA, and the specific EZH2 inhibitor EPZ-6438 (3.3 μM) (PA + EPZ group). EPZ-6438 concentration was chosen according to previous tests. PA time and concentration were chosen based on relevant scientific literature using models and cells similar to our procedure [11–13]. Considering that albumin concentration is essential to determine the concentration of available free fatty acid (FFA), the PA was complexed with bovine serum albumin (BSA; Sigma-Aldrich Cat. No. A9647) at a 4:1 molar ratio considering the albumin concentration already present in the medium due to the FBS supplementation. To warrant the same basal condition, BSA was added in all experimental groups, including the control group (VH).

At 24 h after treatment, cells were lysed for histones extraction as described below, or were fixed in PFA 4% for Oil Red O staining.

2.4. Cell and Tissues Fractionation and Immunoblotting

Cell fractionation was carried out starting with 3×10^6 HUH-7 cells. Briefly, cell pellets were resuspended in buffer A (10 mM Tris, 10 mM KCl, 1.5 mM $MgCl_2$, 300 mM sucrose, and protease inhibitors) and kept on ice for 10 min. For tissue fractionation, 50 mg of liver tissue from each mouse was homogenized in the same buffer A. After centrifugation at 2000× *g* for 10 min, the supernatant contained membranes and the cytosolic fraction (that was subsequently centrifuged at 10,000× *g* for 60 min). Pellets obtained from the first centrifugation were resuspended in buffer B (50 mM Tris 400 mM NaCl, 1 mM EDTA, and 1% NP40) and incubated on ice for 10 min before centrifugation at 5000× *g* for 5 min. The resulting pellet was again resuspended in buffer B plus protease inhibitors by vortexing, incubated on ice, and centrifuged at 5000× *g* for 5 min. The resulting nuclei were pelleted, and protein extraction was carried out by sonication (2 cycles, 1 s). Lysates were then centrifuged at the highest speed for 15 min at 4 °C, and the supernatant was collected (nuclear fraction). Cytosolic and nuclear extracts were then used for immunoblotting. Protein concentration was determined, and proteins were separated by SDS-PAGE under reducing conditions. Following SDS-PAGE, proteins were transferred to nitrocellulose, reacted with specific antibodies, and then detected with peroxidase-conjugated secondary antibodies and a chemiluminescent ECL reagent. Digital images were taken with the Bio-Rad ChemiDoc™ Touch (Hercules, California, USA) and quantified using Bio-Rad Image Lab 5.2.1.

2.5. Oil Red O Cell Staining and Analysis

HUH-7 liver immortalized cells were washed with PBS, stained with a 3:2 Oil Red O solution (0.5% in isopropanol diluted in deionized water; Sigma-Aldrich) for 15 min, and then rinsed.

Slides were scanned with ZEISS Axioscan 7 setting REF_BF H&E Slide 40x program. Quantitative analyses were executed via Qu-Path and ImageJ software. The settings used for analysis were HUE 0–255, saturation 0–255, and brightness 0–185. The stained area was calculated as a percentage of the total area analyzed.

2.6. Statistical Analyses

All statistical analyses and data visualizations were performed in GraphPad Prism 8.0.2. For statistical analysis of body weight and OGTT, two-way ANOVA with the Sidak post-hoc test was used. The other results were analyzed with Student's t-test or one-way ANOVA with Tukey's post-hoc test depending on group number. For all analyses, significance was defined as $p < 0.05$.

3. Results

3.1. An HFD Affects Mice Body Weight and Glucose Homeostasis after 8 Weeks but Not after 2 Weeks

Mice exposed to the hypercaloric diet showed a significant increase in body weight (Figure 1A) starting from the fifth week of diet administration compared to control mice, resulting a full-blown weight gain at 8 weeks (8 W).

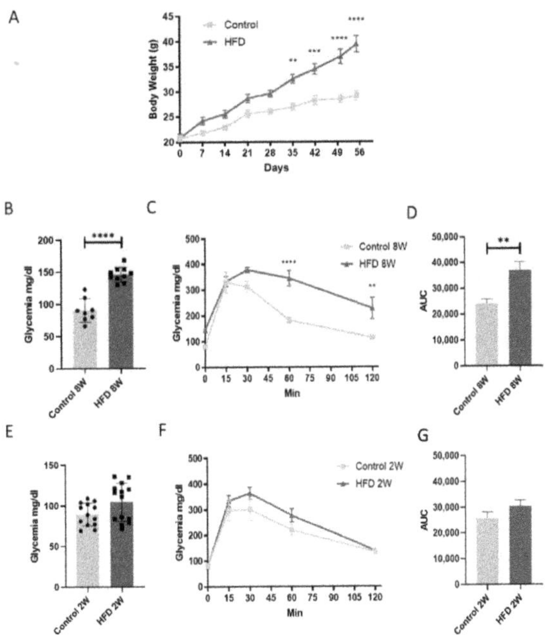

Figure 1. An HFD for 8 W but not 2 W affects mice body weight and glucose homeostasis. (**A**). Mice body weight, glycemia, oral glucose tolerance test, and relative area under the curve (AUC) of mice fed for 8 weeks (8 W) with a control diet or HFD (**B–D**). Glycemia, oral glucose tolerance test, and relative area under the curve (AUC) of mice fed for 2 weeks (2 W) with a control diet or a HFD (**E–G**). ** = $p < 0.01$ vs. control, *** = $p < 0.001$ vs. control, **** = $p < 0.0001$ vs. control. Data are presented as mean ± SEM, n = 8–14 per group.

Weight gain progression correlated with fasting glycemia and OGTT, indicating an impairment in glycemic homeostasis at 8 weeks (Figure 1B–D). Notably, mice fed an HFD for only 2 weeks (2 W) did not exhibit impaired glucose control (Figure 1E–G).

3.2. *An HFD Affects Liver Weight and Steatosis after 8 Weeks but Not after 2 Weeks*

Livers from mice were collected and weighted, and after 8 weeks of an HFD, mice exhibited an increase in liver mass compared to the control group (Figure 2A). In contrast, after just 2 weeks of an HFD, no significant changes in liver mass were observed (Figure 2B).

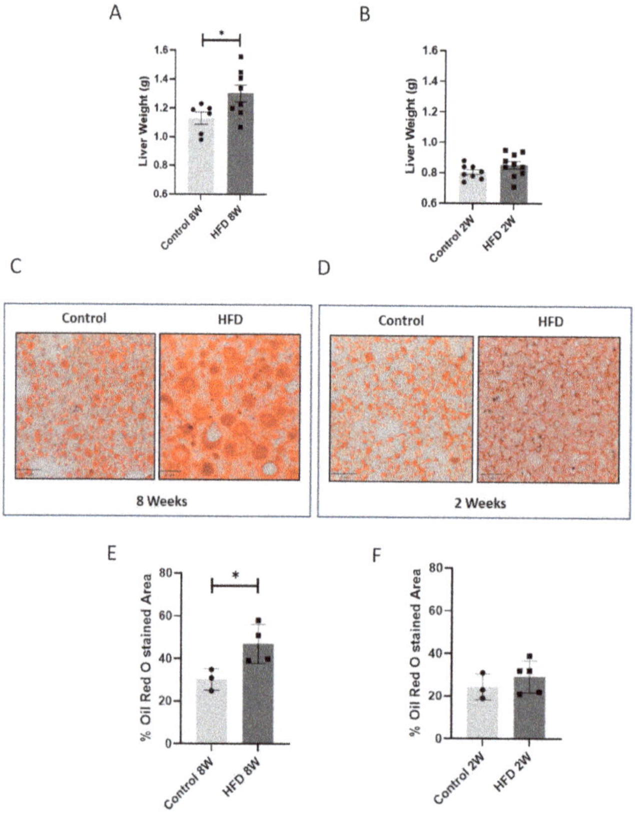

Figure 2. An HFD at 8 W but not 2 W affects mice liver steatosis. Liver weight of mice fed a control diet or an HFD for 8 weeks (**A**) or 2 weeks (**B**). * = $p < 0.05$. Data are presented as mean ± SEM, n = 6–10 per group. Representative images and quantitative analysis of Oil Red O Staining on livers of mice fed a control diet or an HFD for 8 (**C**,**E**) or 2 weeks (**D**,**F**). Line represents a length of 20 μm (**C**,**D**). * = $p < 0.05$. Data are presented as mean ± SEM, n = 3–4 per group.

These results correlated with hepatic steatosis. After 8 weeks of the diet regimen, we observed a marked accumulation of lipid droplets in the HFD 8 W group compared to the control 8 W group (Figure 2C,E), as evidenced by the histological analysis, while no effects were observed in the 2 week HFD-fed animals (Figure 2D,F).

3.3. *Long HFD Exposure Induces an Increase in H3K27me3 Levels Mediated by EZH2*

After establishing that 8 weeks of HFD feeding led to increased body and liver weights (Figures 1A and 2A,B) and impaired glucose homeostasis (Figure 1B–D), we investigated the impact of 8 weeks of HFD exposure on the H3K27me3 (trimethylation of lysine 27 of histone 3) epigenetic modification. Livers obtained from the 8 W control group and 8 W HFD mice were collected, and H3K27me3 levels were analyzed. Histone extraction, followed by Western Blot analysis, revealed that the H3K27me3 profile differed between the two diet regimens. In particular, as shown in Figure 3A, we observed a significant increase ($p < 0.05$) in the prevalence of H3K27me3, without any change in the expression of H3, in response to 8 weeks of HFD feeding. We therefore investigated whether an HFD altered liver levels of H3K27me3 by regulating the expression of EZH2. As reported in Figure 3A, we demonstrated that HFD feeding increased EZH2 expression in the nuclear fraction, while cytoplasmic EZH2 fraction (Figure S2A) was unaffected by HFD feeding. Surprisingly, we did not detect any change in the expression of H3K27me3-associated demethylase KDM6B (Jumonji domain-containing protein-3) (Figure S2B). These data highlight that an in vivo model of HFD feeding is characterized by an intra-hepatic increase in nuclear EZH2, which corresponds to up-regulation of H3K27me3 levels.

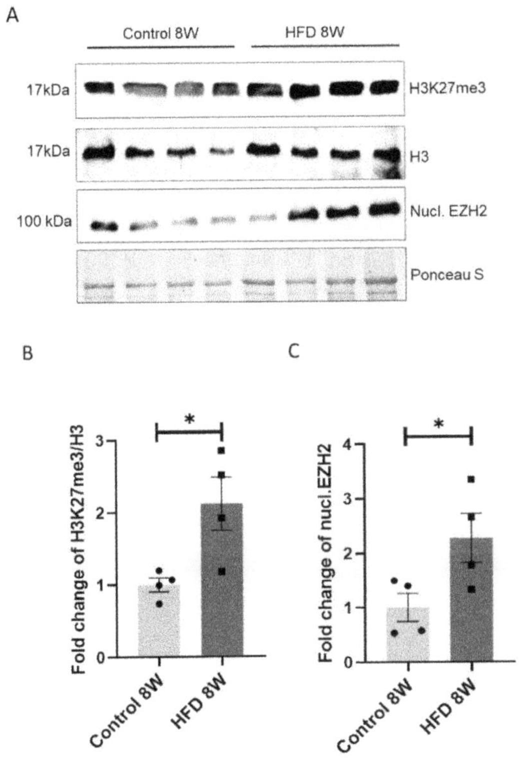

Figure 3. An HFD for 8 W increased H3K27me3 levels and nuclear localization of EZH2. (**A**) Representative Western blotting analysis of nuclear autoradiogram of EZH2 (nucl. EZH2), H3K27me3, and H3 in four mouse liver samples per each group (control and HFD), normalized to total protein (Ponceau S staining). Fold change in (**B**) H3K27me3 levels and (**C**) nuclear EZH2 (nucl. EZH2). * = $p < 0.05$. Data are presented as mean ± SEM, $n = 4$ per group.

3.4. Short HFD Exposure Induces an Increase in H3K27me3 Mediated by EZH2

As reported in Figures 1 and 2, we demonstrated that 8 weeks of an HFD caused an increase in body and liver weight and hepatic steatosis in mice, which correlated with an increase in the prevalence of H3K27me3 in the liver (Figure 3A). The persistence of epigenetic alterations has been described as one of the molecular mechanisms that triggers metabolic memory [14]. Therefore, we extended our investigation to an early time-point, specifically after 2 weeks of HFD feeding. As expected, we observed no effects of an HFD for 2 W on body weight, liver mass, or glucose homeostasis (Figure 1). Upon analyzing the levels of H3K27me3 and EZH2 expression, we observed a significant increase ($p < 0.05$) in H3K27me3 in response to short-term HFD feeding (Figure 4A), with no change in the expression of H3. We therefore investigated the expression levels of EZH2. As reported in Figure 4A, 2 weeks of HFD feeding resulted in increased EZH2 levels in the nuclear fraction, while the cytoplasmic fraction remained unchanged (Figure S2C).

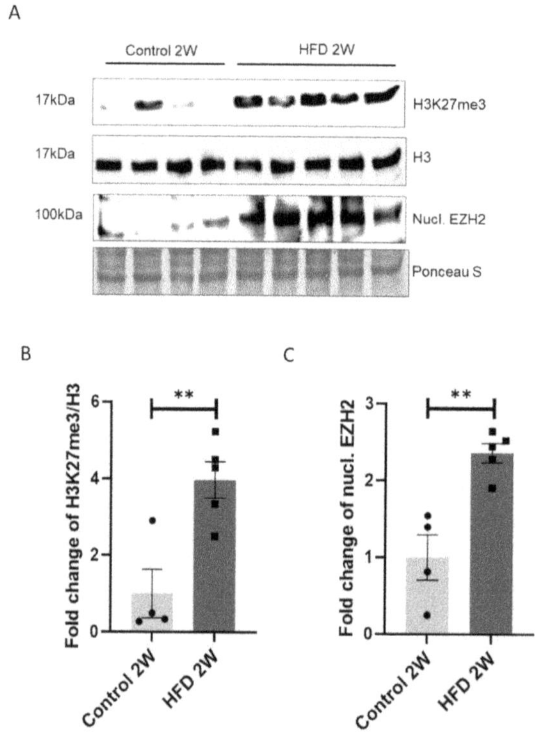

Figure 4. An HFD for 2 W increases H3K27me3 levels and expression and nuclear localization of EZH2. (**A**) Representative Western blot analysis of nuclear level of EZH2 (nucl. EZH2), H3K27me3, and H3 in four mouse liver samples per control group and five per HFD 2 W group, normalized to total protein (Ponceau S staining). Fold change in (**B**) H3K27me3 levels and (**C**) nuclear EZH2 (nucl. EZH2). ** = $p < 0.01$. Data are presented as mean ± SEM, $n = 4/5$ per group.

The reported results indicate that both the upregulation of H3K27me3 and EZH2 occurs in response to HFD exposure, independent of the duration of feeding.

3.5. EPZ-6438, a Selective EZH2 Inhibitor, Reduces H3K27me3 Levels and Counteracts Lipid Accumulation in HUH-7 Cells Treated with Palmitic Acid

To investigate whether pharmacological modulation of H3K27me3 could affect liver lipid accumulation, we established a well-described in vitro model that mimics lipotoxicity induced by free fatty acids. We treated the HUH-7 cells, established from male hepatoma tissue, with palmitic acid (PA) at a concentration of 100 μM for 24 h in the presence or absence of the EZH2 inhibitor EPZ-6438 at 3.3 μM. We investigated whether PA treatment influenced the levels of H3K27me3 and EZH2 expression in both cytoplasmic and nuclear fractions.

As reported in Figure 5A, we demonstrated that PA treatment significantly increased the levels of H3K27me3 and nuclear EZH2, consistent with our findings in mice subjected to HFD feeding. The combined treatment of PA and EPZ-6438 significantly counteracted the induction of H3K27me3 exerted by PA (Figure 5A). Furthermore, we report that PA treatment led to significant accumulation of lipid droplets in HUH-7 cells, as assessed by Oil Red O Staining. Lipid accumulation was partially reversed by the combined treatment of PA and EPZ-6438, as shown in Figure 5D–E. These data clearly demonstrate that the expression and activation of EZH2 play a direct role in lipid accumulation.

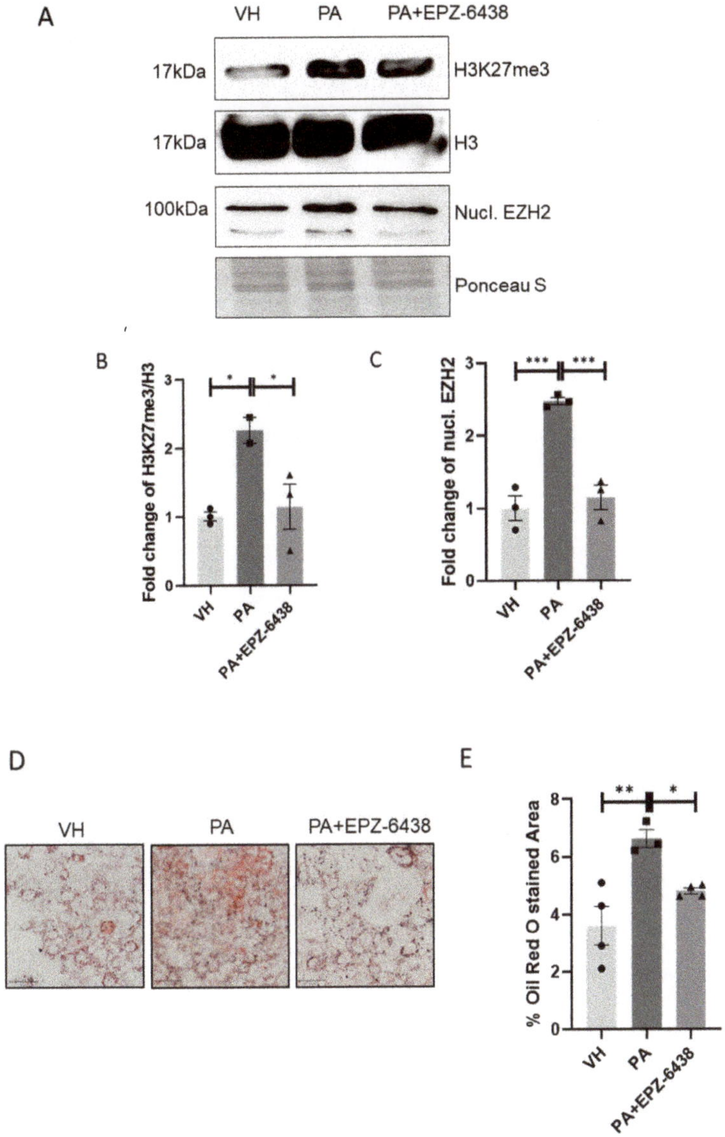

Figure 5. EPZ-6438 treatment reduces H3K27me3 levels and counteracts lipid accumulation induced by palmitic acid in HUH-7 cells. (**A**) Representative Western blot analysis of nuclear EZH2 (nucl. EZH2), H3K27me3, and H3 in HUH cells treated for 24 h with 100 μM palmitic acid in combination or not with EPZ-6438 3.3 μM. Fold change of (**B**) H3K27me3 levels and (**C**) nuclear EZH2 (nucl. EZH2). Representative images and quantitative analysis of Oil Red O staining on HUH-7 cells treated with PA with or without EPZ-6438 (**D**,**E**). Line represents a length of 50 μm (**D**). * $p < 0.05$, ** $p < 0.01$, *** $p < 0.001$. Data are presented as mean ± SEM, n = 3–4 experiments.

4. Discussion

Overweight and obesity are significant risk factors for lipid accumulation in the liver. Excessive hepatic fat accumulation can lead to inflammation, liver cell injury, and the progression to more severe conditions such as non-alcoholic steatohepatitis (NASH) and, eventually, cirrhosis [15,16]. Extensive research indicates that the intricate interplay of multiple factors, including insulin resistance, inflammation, and oxidative stress, contributes to the pathogenesis of hepatic lipid accumulation in individuals who are overweight or obese. Of note, the initial detrimental mechanisms leading to fat accumulation in the liver begin to onset even at early stages of metabolic derangement and even before the onset of overweight [17]. The factors involved in the early/late onset of liver-associated metabolic derangement have not yet been fully investigated, but a substantial body of evidence suggests that epigenetic mechanisms could play a pivotal role.

Among epigenetic players, EZH2, through H3K27me3, is known to exert a crucial role in the regulation of inflammatory and fibrotic markers in liver parenchyma [9,18]. Data reported by Mann and colleagues revealed that the expression of EZH2 was up regulated in activated hepatic stellate cells (HSCs) compared with quiescent HSCs [19]. Emerging evidence indicates HSCs exert a pivotal role in hepatic fibrosis [19], a typical consequence of chronic diet-induced dysmetabolism [20]. EZH2 may drive hepatic fibrosis via H3K27me3 in order to repress anti-fibrotic gene expression [21]. Among those under H3K27me3 control, small mother against decapentaplegic 7 (Smad7) and Dickkopf-related protein 1 (Dkk1) are two anti-fibrotic genes known for their ability to inhibit pro-fibrotic signaling mediated by TGF-β [22,23] and Wnt [24,25], respectively. Treatment with an EZH2 inhibitor, indeed, could prevent the trans differentiation of HSCs to highly proliferative and fibrogenic myofibroblast-like cells (MFBs) [26]. Some authors reported that EZH2 silencing restored Dkk1 expression, preventing TGF-β-induced proliferation of HSCs and expression of alpha smooth muscle Actin (α-SMA) [26].

In line with these data, Rosa Martin-Mateos and colleagues demonstrated that EZH2 is strongly involved in TGF-β-mediated HSC activation. Inhibition of EZH2 attenuates fibrogenic gene transcription in TGF-β-mediated HSC activation in in vitro and in vivo liver fibrosis murine models, reducing the expression of fibronectin, α-SMA, and collagen 1α1 [27]. Additionally, the EZH2 inhibitor promoted the expression of the activin membrane-bound inhibitor (BAMBI); IL-10 (interleukin-10); and cell cycle regulators, including cyclin-dependent kinase inhibitor 1A (Cdkn1a) as well as growth arrest and DNA damage-inducible alpha (Gadd45a) and beta (Gadd45b), leading to TGF-β/Smads suppression and an anti-inflammatory response [26]. Additionally, another anti-fibrotic gene repressed by EZH2 is peroxisome proliferator-activated receptor gamma (PPAR-γ) [28], a ligand-activated transcription factor that promotes protective effects against including liver fibrosis [29].

In this study, we demonstrate an increase in EZH2 expression and nuclear localization as well as increased H3K27me3 levels in the livers of overweight mice that were fed an HFD for 8 weeks. Moreover, we did not observe any changes in the levels of the H3K27me3-specific demethylase KDM6B. H3K27me3 is known for its long-lasting effects on transcriptional regulation in conditions of diet-induced metabolic derangement. In this context, Blin and colleagues demonstrated that chromatin-repressive marks such as H3K27me3 represent a potential candidate in the understanding of the long-term "developmental programming" effects induced by early exposure to nutritional excess [30]. Thereby, EZH2-mediated H3K27me3 can be considered among the players involved in the so-called "metabolic memory". The hypothesis of metabolic memory suggests that uncontrolled glycemic/weight control in early stages of metabolic derangement can have a long-lasting effect on increasing the risk of complication onset later in life [31]. The mechanisms involved in metabolic memory have been so far poorly investigated; epigenetic modifications, and particularly H3K27me3, could represent a promising area for further investigation in this context.

To understand how early H3K27me3 occurs, we continued our experimental protocol analyzing the livers of mice fed an HFD for only 2 weeks (short HFD group), which corresponds to a time that precedes the onset of overweight and the impairment of glucose homeostasis. Surprisingly, we detected, at this early time-point, a significant upregulation of the nuclear levels of EZH2 and H3K27me3, similar to what we observed in overweight mice (HFD for 8 weeks). Here we reveal, for the first time, an early epigenetic alteration resulting from a short HFD challenge occurring prior to the onset of a full-blown overweight/glucose impairment. These findings contribute to consolidating the hypothesis concerning the involvement of EZH2/H3K27me3 in metabolic memory. To strengthen this idea, we plan to perform a new in vivo experimental protocol consisting of a short HFD priming of 2 weeks (similar to the current study), followed by a recovery phase on a control diet, and then a new HFD challenge.

To properly correlate the hepatic modulation of EZH2 expression upon short/long HFD consumption, we established a well-described in vitro model of non-alcoholic fatty liver disease (NAFLD) using immortalized HUH-7 cells challenged with PA (24 h, 100 μM) [32] and EPZ-6438 (24 h, 3.3 μM).

Firstly, we demonstrated that the results obtained from the in vitro model after PA challenge mimicked those obtained in vivo after an HFD. Specifically, we observed an increase in the expression of EZH2/H3K27me3 in cells that received PA compared to vehicle-treated cells. As a readout for hepatic metabolic condition, we measured liver weight in vivo, while we evaluated the accumulation of intracellular lipid droplets in vitro. As predicted, PA treatment induced an increase in fat droplets.

EPZ-6438 treatment reduced H3K27me3 levels and nuclear expression/localization of EZH2 induced by PA. More interestingly, it significantly counteracted PA-dependent lipid accumulation.

These results clearly demonstrate that EZH2 serves as a direct regulator of fat accumulation. Very recently, EZH2 has been proposed as a potential therapeutic target for advanced NAFLD [9]. Our results reveal that EZH2 deregulation is an early indicator of metabolic derangement. For this reason, we believe that EZH2-targeted therapy could represent an attractive early approach to mitigate the late risk of liver injury. Currently, EPZ-6438 (tazemetostat) is a first-in-class U.S. FDA-approved EZH2 inhibitor for follicular lymphoma and epithelioid sarcoma [33–35].

Even in the context of drug repurposing, it will be strategic to perform an in vivo pharmacological study with EPZ-6438 in the setting of both short and long HFD feeding. Finally, our next goal to overcome the limitation of this study will be to investigate the gene sets regulated by H3K27me3 through RNA sequencing followed by CHIP (chromatin immune precipitation) analysis with the aim of identifying more specific druggable molecular pathways.

5. Conclusions

In conclusion, we identify EZH2-mediated H3K27me3 as an early epigenetic change occurring in livers challenged by fatty acids, both in vivo and in vitro. Our finding supports the hypothesis that EZH2 represents a potential pharmacological target for addressing metabolic disorders.

Supplementary Materials: The following supporting information can be downloaded at: https://www.mdpi.com/article/10.3390/nu16193260/s1, Figure S1: Schematic representation of in vivo experimental dietary protocol. Figure S2: An HFD for 8 W and 2 W did not regulate cytosolic EZH2. An HFD for 8 W did not change KDM6B expression.

Author Contributions: Conceptualization, G.P. and F.C.; methodology, G.P., M.P., V.G., E.M., N.C., J.M., G.C., G.D. and F.C.; formal analysis, G.P, M.P., E.M. and F.C.; writing—original draft preparation, G.P and F.C.; writing—review and editing, G.P., D.L., L.M. and F.C.; supervision, F.C.; funding acquisition, G.P. All authors have read and agreed to the published version of the manuscript.

Funding: This research was funded by Bando Ricerca UPO 2022, (MUR-DM 737/2021 of PNR and Compagnia di San Paolo).

Institutional Review Board Statement: The animal study protocol was in accordance with the Italian law on animal care (D.L. 26/2014) as well as the European Directive (2010/63/UE) and approved by the Organismo Preposto al Benessere Animale (OPBA) of University of Piemonte Orientale, Novara, Italy (code: DB064.61; date: 15 October 2020).

Informed Consent Statement: Not applicable.

Data Availability Statement: The raw data supporting the conclusions of this article will be made available by the authors on request.

Conflicts of Interest: The authors declare no conflicts of interest.

References

1. Chiazza, F.; Challa, T.D.; Lucchini, F.C.; Konrad, D.; Wueest, S. A Short Bout of HFD Promotes Long-Lasting Hepatic Lipid Accumulation. *Adipocyte* **2016**, *5*, 88–92. [CrossRef] [PubMed]
2. Benetti, E.; Mastrocola, R.; Vitarelli, G.; Cutrin, J.C.; Nigro, D.; Chiazza, F.; Mayoux, E.; Collino, M.; Fantozzi, R. Empagliflozin Protects against Diet-Induced NLRP-3 Inflammasome Activation and Lipid Accumulation. *J. Pharmacol. Exp. Ther.* **2016**, *359*, 45–53. [CrossRef]
3. Gopinath, V.; Mariya Davis, A.; Menon, T.K.; Raghavamenon, A.C. Alcohol Promotes Liver Fibrosis in High Fat Diet Induced Diabetic Rats. *J. Basic Clin. Physiol. Pharmacol.* **2024**. [CrossRef] [PubMed]
4. González-Becerra, K.; Ramos-Lopez, O.; Barrón-Cabrera, E.; Riezu-Boj, J.I.; Milagro, F.I.; Martínez-López, E.; Martínez, J.A. Fatty Acids, Epigenetic Mechanisms and Chronic Diseases: A Systematic Review. *Lipids Health Dis.* **2019**, *18*, 178. [CrossRef] [PubMed]
5. Mahmoud, A.M. An Overview of Epigenetics in Obesity: The Role of Lifestyle and Therapeutic Interventions. *Int. J. Mol. Sci.* **2022**, *23*, 1341. [CrossRef] [PubMed]
6. Duan, R.; Du, W.; Guo, W. EZH2: A Novel Target for Cancer Treatment. *J. Hematol. Oncol.* **2020**, *13*, 104. [CrossRef]
7. Skourti-Stathaki, K.; Torlai Triglia, E.; Warburton, M.; Voigt, P.; Bird, A.; Pombo, A. R-Loops Enhance Polycomb Repression at a Subset of Developmental Regulator Genes. *Mol. Cell* **2019**, *73*, 930–945.e4. [CrossRef]
8. Le, H.Q.; Hill, M.A.; Kollak, I.; Keck, M.; Schroeder, V.; Wirth, J.; Skronska-Wasek, W.; Schruf, E.; Strobel, B.; Stahl, H.; et al. An EZH2-dependent Transcriptional Complex Promotes Aberrant Epithelial Remodelling after Injury. *EMBO Rep.* **2021**, *22*, e52785. [CrossRef]
9. Lim, H.J.; Kim, M. EZH2 as a Potential Target for NAFLD Therapy. *Int. J. Mol. Sci.* **2020**, *21*, 8617. [CrossRef]
10. Pellizzon, M.A.; Ricci, M.R. The Common Use of Improper Control Diets in Diet-Induced Metabolic Disease Research Confounds Data Interpretation: The Fiber Factor. *Nutr. Metab.* **2018**, *15*, 3. [CrossRef]
11. Osorio, D.; Pinzón, A.; Martín-Jiménez, C.; Barreto, G.E.; González, J. Multiple Pathways Involved in Palmitic Acid-Induced Toxicity: A System Biology Approach. *Front. Neurosci.* **2019**, *13*, 1410. [CrossRef] [PubMed]
12. Ruiz, M.; Henricsson, M.; Borén, J.; Pilon, M. Palmitic Acid Causes Increased Dihydroceramide Levels When Desaturase Expression Is Directly Silenced or Indirectly Lowered by Silencing AdipoR2. *Lipids Health Dis.* **2021**, *20*, 173. [CrossRef] [PubMed]
13. Maruyama, H.; Takahashi, M.; Sekimoto, T.; Shimada, T.; Yokosuka, O. Linoleate Appears to Protect against Palmitate-Induced Inflammation in Huh7 Cells. *Lipids Health Dis.* **2014**, *13*, 78. [CrossRef] [PubMed]
14. Asif, S.; Morrow, N.M.; Mulvihill, E.E.; Kim, K.-H. Understanding Dietary Intervention-Mediated Epigenetic Modifications in Metabolic Diseases. *Front. Genet.* **2020**, *11*, 590369. [CrossRef]
15. Parlati, L.; Régnier, M.; Guillou, H.; Postic, C. New Targets for NAFLD. *JHEP Rep.* **2021**, *3*, 100346. [CrossRef]
16. Geng, Y.; Faber, K.N.; de Meijer, V.E.; Blokzijl, H.; Moshage, H. How Does Hepatic Lipid Accumulation Lead to Lipotoxicity in Non-Alcoholic Fatty Liver Disease? *Hepatol. Int.* **2021**, *15*, 21–35. [CrossRef]
17. Ipsen, D.H.; Lykkesfeldt, J.; Tveden-Nyborg, P. Molecular Mechanisms of Hepatic Lipid Accumulation in Non-Alcoholic Fatty Liver Disease. *Cell. Mol. Life Sci.* **2018**, *75*, 3313–3327. [CrossRef]
18. Lee, S.; Woo, D.-C.; Kang, J.; Ra, M.; Kim, K.H.; Lee, S.R.; Choi, D.K.; Lee, H.; Hong, K.B.; Min, S.-H.; et al. The Role of the Histone Methyltransferase EZH2 in Liver Inflammation and Fibrosis in STAM NASH Mice. *Biology* **2020**, *9*, 93. [CrossRef]
19. Mann, J.; Chu, D.C.K.; Maxwell, A.; Oakley, F.; Zhu, N.-L.; Tsukamoto, H.; Mann, D.A. MeCP2 Controls an Epigenetic Pathway That Promotes Myofibroblast Transdifferentiation and Fibrosis. *Gastroenterology* **2010**, *138*, 705–714. [CrossRef]
20. Kandhi, R.; Menendez, A.; Ramanathan, S.; Ilangumaran, S. Regulation of High-Fat Diet-Induced Liver Fibrosis by SOCS1 Expression in Hepatic Stellate Cells. *J. Clin. Exp. Hepatol.* **2024**, *14*, 101280. [CrossRef]
21. Basta, M.D.; Petruk, S.; Mazo, A.; Walker, J.L. Fibrosis-the Tale of H3K27 Histone Methyltransferases and Demethylases. *Front. Cell Dev. Biol.* **2023**, *11*, 1193344. [CrossRef] [PubMed]
22. Biernacka, A.; Dobaczewski, M.; Frangogiannis, N.G. TGF-β Signaling in Fibrosis. *Growth Factors* **2011**, *29*, 196–202. [CrossRef] [PubMed]
23. Meng, X.-M.; Nikolic-Paterson, D.J.; Lan, H.Y. TGF-β: The Master Regulator of Fibrosis. *Nat. Rev. Nephrol.* **2016**, *12*, 325–338. [CrossRef] [PubMed]

24. Akhmetshina, A.; Palumbo, K.; Dees, C.; Bergmann, C.; Venalis, P.; Zerr, P.; Horn, A.; Kireva, T.; Beyer, C.; Zwerina, J.; et al. Activation of Canonical Wnt Signalling Is Required for TGF-β-Mediated Fibrosis. *Nat. Commun.* **2012**, *3*, 735. [CrossRef] [PubMed]
25. Wang, J.-N.; Li, L.; Li, L.-Y.; Yan, Q.; Li, J.; Xu, T. Emerging Role and Therapeutic Implication of Wnt Signaling Pathways in Liver Fibrosis. *Gene* **2018**, *674*, 57–69. [CrossRef]
26. Jiang, Y.; Xiang, C.; Zhong, F.; Zhang, Y.; Wang, L.; Zhao, Y.; Wang, J.; Ding, C.; Jin, L.; He, F.; et al. Histone H3K27 Methyltransferase EZH2 and Demethylase JMJD3 Regulate Hepatic Stellate Cells Activation and Liver Fibrosis. *Theranostics* **2021**, *11*, 361–378. [CrossRef]
27. Martin-Mateos, R.; De Assuncao, T.M.; Arab, J.P.; Jalan-Sakrikar, N.; Yaqoob, U.; Greuter, T.; Verma, V.K.; Mathison, A.J.; Cao, S.; Lomberk, G.; et al. Enhancer of Zeste Homologue 2 Inhibition Attenuates TGF-β Dependent Hepatic Stellate Cell Activation and Liver Fibrosis. *Cell Mol. Gastroenterol. Hepatol.* **2019**, *7*, 197–209. [CrossRef]
28. Deng, Y.-L.; Xiong, X.-Z.; Cheng, N.-S. Organ Fibrosis Inhibited by Blocking Transforming Growth Factor-β Signaling via Peroxisome Proliferator-Activated Receptor γ Agonists. *Hepatobiliary Pancreat. Dis. Int.* **2012**, *11*, 467–478. [CrossRef]
29. Li, J.; Guo, C.; Wu, J. The Agonists of Peroxisome Proliferator-Activated Receptor-γ for Liver Fibrosis. *Drug Des. Dev. Ther.* **2021**, *15*, 2619–2628. [CrossRef]
30. Blin, G.; Liand, M.; Mauduit, C.; Chehade, H.; Benahmed, M.; Simeoni, U.; Siddeek, B. Maternal Exposure to High-Fat Diet Induces Long-Term Derepressive Chromatin Marks in the Heart. *Nutrients* **2020**, *12*, 181. [CrossRef]
31. Dong, H.; Sun, Y.; Nie, L.; Cui, A.; Zhao, P.; Leung, W.K.; Wang, Q. Metabolic Memory: Mechanisms and Diseases. *Signal Transduct. Target. Ther.* **2024**, *9*, 38. [CrossRef] [PubMed]
32. Ricchi, M.; Odoardi, M.R.; Carulli, L.; Anzivino, C.; Ballestri, S.; Pinetti, A.; Fantoni, L.I.; Marra, F.; Bertolotti, M.; Banni, S.; et al. Differential Effect of Oleic and Palmitic Acid on Lipid Accumulation and Apoptosis in Cultured Hepatocytes. *J. Gastroenterol. Hepatol.* **2009**, *24*, 830–840. [CrossRef] [PubMed]
33. Gounder, M.; Schöffski, P.; Jones, R.L.; Agulnik, M.; Cote, G.M.; Villalobos, V.M.; Attia, S.; Chugh, R.; Chen, T.W.-W.; Jahan, T.; et al. Tazemetostat in Advanced Epithelioid Sarcoma with Loss of INI1/SMARCB1: An International, Open-Label, Phase 2 Basket Study. *Lancet Oncol.* **2020**, *21*, 1423–1432. [CrossRef] [PubMed]
34. Morschhauser, F.; Tilly, H.; Chaidos, A.; McKay, P.; Phillips, T.; Assouline, S.; Batlevi, C.L.; Campbell, P.; Ribrag, V.; Damaj, G.L.; et al. Tazemetostat for Patients with Relapsed or Refractory Follicular Lymphoma: An Open-Label, Single-Arm, Multicentre, Phase 2 Trial. *Lancet Oncol.* **2020**, *21*, 1433–1442. [CrossRef]
35. Groisberg, R.; Subbiah, V. EZH2 Inhibition for Epithelioid Sarcoma and Follicular Lymphoma. *Lancet Oncol.* **2020**, *21*, 1388–1390. [CrossRef]

nutrients

MDPI

Article

Comprehensive Analysis of *Berberis aristata* DC. Bark Extracts: In Vitro and In Silico Evaluation of Bioaccessibility and Safety

Giovanna Rigillo [1,2,*], Giorgio Cappellucci [2,3], Giulia Baini [2,3], Federica Vaccaro [2,3], Elisabetta Miraldi [2,3], Luca Pani [1,4], Fabio Tascedda [5,6], Renato Bruni [7] and Marco Biagi [2,7]

1 Department of Biomedical, Metabolic and Neural Sciences, University of Modena and Reggio Emilia, 41125 Modena, Italy; luca.pani@unimore.it
2 Laboratory of Italian Society of Phytoterapy-SIFITLab, 53100 Siena, Italy; giorgi.cappellucci@unisi.it (G.C.); giulia.baini2@unisi.it (G.B.); fvaccaro@student.unisi.it (F.V.); elisabetta.miraldi@unisi.it (E.M.); marco.biagi@unipr.it (M.B.)
3 Department of Physical Sciences, Earth and Environment, University of Siena, 53100 Siena, Italy
4 Department of Psychiatry and Behavioral Sciences, University of Miami, Miami, FL 33136, USA
5 Department of Life Sciences, University of Modena and Reggio Emilia, 41125 Modena, Italy; fabio.tascedda@unimore.it
6 Consorzio Interuniversitario Biotecnologie (CIB), 34148 Trieste, Italy
7 Department of Food and Drug, University of Parma, 43124 Parma, Italy; renato.bruni@unipr.it
* Correspondence: giovanna.rigillo@unimore.it

Abstract: Berberine (BER) is an alkaloid found, together with other protoberberinoids (PROTBERs), in several species used in medicines and food supplements. While some herbal preparations containing BER and PROTBERs, such as *Berberis aristata* DC. bark extracts, have shown promising potential for human health, their safety has not been fully assessed. Recently, the EFSA issued a call for data to deepen the pharmacokinetic and pharmacodynamic understanding of products containing BER and PROTBERs and to comprehensively assess their safety, especially when used in food supplements. In this context, new data were collected in this work by assessing: (i) the phytochemical profile of 16 different commercial *B. aristata* dry extracts, which are among the most widely used preparations containing BER and PROTBERs in Europe; (ii) the In Vitro and In Silico investigation of the pharmacokinetic properties of BER and PROTBERs; (iii) the In Vitro cytotoxicity of selected extracts in different human cell lines, including tests on hepatic cells in the presence of CYP450 substrates; (iv) the effects of the extracts on cancer cell migration; and (v) the In Vitro molecular effects of extracts in non-cancer human cells. Results showed that commercial *B. aristata* extracts contain BER as the main constituent, with jatrorrhizine as main secondary PROTBER. BER and jatrorrhizine were found to have a good bioaccessibility rate, but they interact with P-gp. *B. aristata* extracts showed limited cytotoxicity and minimal interaction with CYP450 substrates. Furthermore, tested extracts demonstrated inhibition of cancer cell migration and were devoid of any pro-tumoral effects in normal cells. Overall, our work provides a valuable overview to better elucidate important concerns regarding botanicals containing BER and PROTBERs.

Keywords: berberine; safety; bioaccessibility; cell viability; gene expression

1. Introduction

Berberine (BER) is an isoquinoline alkaloid belonging to the category of protoberberines, including berberrubine, thalifendine, demethyleneberberine, coptisine, jatrorrhizine, columbamine, palmatine, and epiberberine. These compounds share a common molecular moiety, characterized by an aromatic quaternary ammonium nitrogen, a bright yellow color, and water solubility [1] (Figure 1).

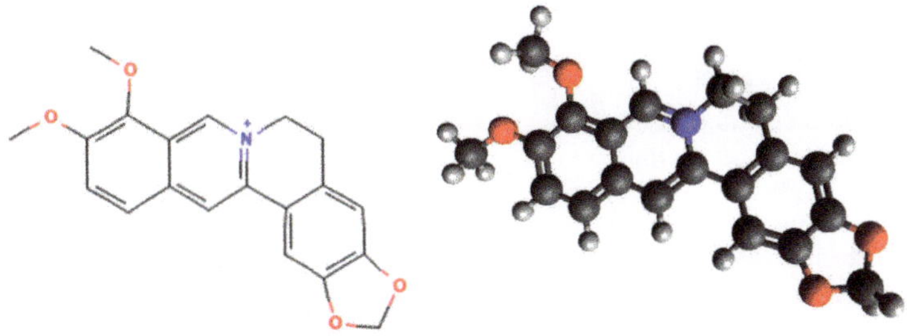

Figure 1. Berberine chemical structure in 2D and 3D (PubChem CID: 2353) (https://pubchem.ncbi.nlm.nih.gov/compound/Berberine, accessed on 22 August 2024).

BER is a characteristic compound with biological activity found in several plant species with a well-established use in traditional Asian medicine, such as those in *Berberis*, *Coptis*, *Phellodendron*, *Corydalis*, and *Xanthorhiza* genera or in *Tinospora sinensis* (Lour.) Merr. and *Coscininium fenestratum* (Goetgh.) Colebr., but it also occurs in medicinal species known in American and European Western medicine, such as *Hydrastis canadensis* L., *Chelidonium majus* L., or *Papaver* spp., as well as in Siberia (*Thalictrum flavum* L.) and in Africa (*Jateorhiza palmata* (Lam.) Miers and *Annickia chlorantha* (Oliv.) Setten & Maas) [2]. In herbal preparations obtained from these species, other compounds may contribute to the biological activity of BER, in particular the alkaloids chemically related to BER, such as protoberberinoid derivatives (PROTBERs) [3].

To date, in the European Union (EU), an official monograph on the medicinal use of these species has not been produced by the European Medicines Agency (EMA), while a *Berberis aristata* DC. stem monograph was added to the European Pharmacopoeia in 2022, being the most commonly used BER-containing species used in food supplements [4,5].

BER and BER-containing herbal preparations from *B. aristata* have displayed a plethora of biological effects, as witnessed by numerous clinical trials. A high-quality meta-analysis [3] reports that most clinical trials concern cardiovascular system disorders (atherosclerosis and lipid profile) [5–8], inflammation [9,10], glycemia [11–14], gastrointestinal health [15,16], and cancer [17–19]. Evidence indicates that BER shows beneficial properties in conditions of dysregulation or alteration of physiological processes, displaying various activities such as lipid and glycemic regulation, anti-inflammatory, antioxidant and antiproliferative activities, and gut microbiota modulation [20]. The biological activity of BER has been associated with multiple molecular mechanisms involving different cell signaling pathways; PKA, p38 MAPK, Wnt/β-catenin, AMPK, RANK/RANKL/OPG, PI3K/Akt, NFAT, NF-κB, Hedgehog, low-density lipoprotein (LDL) receptor expression, reactive oxygen species (ROS), and nitric oxide (NO) production have been described as molecular targets of BER in different conditions regarding sugars and lipids metabolism, cardiovascular, gastrointestinal, musculoskeletal, and the central nervous systems, as well as cancer and inflammation [21–26].

Potential adverse events could be linked to the biological activity of BER and PROTBERs; therefore, the safety of BER, PROTBERs, and BER-containing herbal preparations has been carefully considered.

In terms of pharmacokinetics, BER and PROTBERs show different bioavailability: BER is a substrate for P-glycoprotein (P-gp) and less than 1% of an oral dose was found to be bioavailable [27]; also, jatrorrhizine is a P-gp substrate and it showed a poor bioavailability [28,29]. Other PROTBERs have a higher bioavailability, as demonstrated by in vivo studies for epiberberine [30], palmatine [31], berberubbine [32], coptisine, and columbamine [33]. Despite the lack of proper knowledge on cellular bioavailability, it is well known that after intestinal absorption, BER and PROTBERs are rapidly metabolized by hepatic CYP450 enzymes. Many studies have primarily described BER metabolism, reporting that the CYP450 1A2, 3A4, 2D6, and 2C9 isoforms are mainly involved [34], while few have studied other PROTBERs and their interaction with different CYP450 isoforms, with the exception of jatrorrhizine, metabolized by CYP1A2 [35], and palmatine by CYP2D6 [36]. It is plausible that CYP1A2, CYP3A4, and CYP2D6 have a key role in the metabolism of many PROTBERs; thus, drug–herbal product interactions should be considered when BER, PROTBERs, and BER-containing herbal preparations are used.

Given the complexity of the topic and the well-established and growing consumption of BER-containing herbal preparations, especially in food supplements, in 2023 the European Food Safety Authority (EFSA) opened a "call for data" to obtain relevant updates on the safety assessment of BER-containing herbal preparations used in food supplements [37].

Within this context, this study aims to provide new experimental-based evidence to fill some pivotal gaps concerning the chemical and pharmacokinetic aspects, and biological safety of berberine. To accomplish our aim, this research was structured into seven steps:

(1) the phytochemical characterization of sixteen different commercial raw BER-containing ingredients, used in food supplement formulation, through the development of an efficient HPLC-DAD method to quantify BER and PROTBERs;
(2) In Vitro evaluation of digestive stability and bioaccessibility, and In Silico investigation of pharmacokinetic properties of BER and PROTBERs;
(3) prediction of potential BER-related targets by bioinformatic analysis;
(4) assessment of cytotoxicity in In Vitro human intestinal (Caco-2), hepatic (HepG2), gastric (AGS), and kidney (HEK293) cell lines, and of BER's impact on hepatic cell viability in the presence of CYP450 substrates;
(5) evaluation of BER's effects on cell migration ability of colorectal carcinoma cells (Caco-2);
(6) investigation of potential oxidative stress of BER in non-tumoral kidney cells (HEK293) by the dosage of ROS;
(7) evaluation of the transcriptional effects of BER on the main target genes involved in the regulation of cell cycle, cell growth, and neoplastic transformation, as well as on oncogenes.

2. Materials and Methods

2.1. Plant Material

Sixteen different extracts used as raw ingredients in Italy for food supplement preparations were selected from the international business-to-business market. According to their declared content, they were divided into two groups: eight *B. aristata* bark extracts with a declared content of 97% BER *w*/*w* dry basis (db) as BER hydrochloride, and the remnant with a declared content of 85% BER *w*/*w* db. Samples were kindly supplied by EHPM (Brussels, Belgium), Viatris (Monza-Brianza, Italy), Giellepi (Seregno, Monza-Brianza, Italy), and Vivatis (Gallarate, Varese, Italy).

2.2. Phytochemical Analysis

The quantification of BER and PROTBERs in the examined extracts was determined by analytical techniques and measured by calculating values as BER hydrochloride on the dry basis (db) of the extract (% *w*/*w* db). All samples were dissolved in ultra-pure water (1 mg/mL) and analyzed by using an HPLC-DAD Shimadzu Prominence LC 2030 3D instrument. A Bondpak® C18 column, 10 µm, 125 Å, 3.9 mm × 300 mm (Waters Corporation, Milford, MA, USA), was used as the stationary phase. The mobile phase was composed of water with 0.1% *v*/*v* formic acid (A) and acetonitrile with 0.1% *v*/*v* formic acid (B). The following method was used: B from 45% at 0 min and a linear increase to 55% at 9 min, then 45% at 10 min, holding the same percentage until the end, 11 min. Flow rate was set at 0.900 mL/min; column temperature was 30 °C. The chromatogram was recorded at 346 nm. BER hydrochloride (Merck KGaA, Darmstadt, Germany) (10–0.01 µg in column) was used to build the calibration curve. In herbal preparations, BER was identified and quantified by using the specific external standard, whereas other PROTBERs were identified according to the literature and their UV–vis spectra [31]; PROTBERs were quantified and expressed as BER hydrochloride.

Given the homogeneous phytochemical profile of the extracts, subsequent bioactivities were determined using two different *B. aristata* bark extracts with a declared content of 97% BER *w*/*w*, and two *B. aristata* bark extracts with a BER content of 85% *w*/*w*, namely, A85, D85, A97, and G97 (Table 3 in Section 3.1).

2.3. *In Vitro Bioaccessibility Assessment*

The test was conducted according to Governa et al., 2022 [38] and the validated INFOGEST protocol [39], with some modifications. In detail, 20 mg of each extract were added to 20 mL of simulated gastric juice, containing pepsin from porcine gastric mucosa (300 UI/mL, Merck) and NaCl (10 mg/mL); the pH of the solution was adjusted to 1.7 using HCl. Samples were incubated for 2 h at 37 °C with shaking. Then, pancreatin from porcine pancreas (activity equivalent to 4× U.S.P., 10 mg/mL, Merck) and a bile salt mixture (20 mg/mL, Merck) were added, and the pH was increased to 7.0 by adding $NaHCO_3$ (15 mg/mL, Merck) to simulate the intestinal environment. Intestinal digestion was carried out for 2 h at 37 °C with shaking. Samples were then filtered and immediately used for further analysis, performed according to the HPLC-DAD method described above. The bioaccessibility rate of each compound was calculated as the % of its recovery after digestion, compared to the initial amount. Two independent experiments were performed.

2.4. *In Silico Pharmacokinetic Analysis and Target Prediction*

Computational analysis of pharmacokinetic characteristics of BER and PROTBERs present in the most common food supplements marketed in the EU was performed using the SwissADME© web tool, a free platform created and developed by the Molecular Modeling Group of the SIB, Swiss Institute of Bioinformatics. According to their chemical–physical characteristics, compound similarities, and provided data, the trained algorithm estimates compounds for ADME (absorption, distribution, metabolism, and excretion) properties, physical chemistry, drug likeness, pharmacokinetics, and medicinal chemistry properties [40]. Target prediction was performed by using GeneCards® suite, a web tool developed to obtain results with functionality and relevance scoring, allowing for the combination of query terms and providing relevant literature, which the match is based on [41]. In this work, SwissTargetPrediction, another web tool of the SIB, was used, as well as SEA (Similarity Ensemble Approach). SEA predicts the biological targets of a compound based on its resemblance to ligands annotated in reference databases and relates proteins by their pharmacology by aggregating chemical similarity among entire sets of ligands; SEA also scores Tanimoto similarity calculations based on compound annotations derived from ChEMBL [42].

2.5. *In Vitro Cell Culture and Treatment*

Human colorectal (Caco-2), hepatic (HepG2), gastric (AGS), and kidney (HEK293) cell lines were cultured in DMEM supplemented with 10% fetal bovine serum (FBS), 1% glutamine, and 1% penicillin/streptomycin antibiotics. Media and material for cell cultures was supplied by Merck. Cells were maintained under a humidified atmosphere of 5% CO_2 at 37 °C. Cells were treated with two representative samples based on the results obtained from chemical analysis: one *B. aristata* bark extract with 97% BER (as berberine hydrochloride) *w*/*w*, composed of two pooled extracts containing BER 97% (B97%) and one *B. aristata* bark extracts with 85% BER content *w*/*w*, composed of two pooled extracts containing BER 85% (B85%). The control group received phosphate-buffered saline (PBS). The treatments were carried out at various time points, as indicated in each section, for further analysis.

2.6. *Cytotoxicity Assay*

Caco-2, HepG2, HEK293, and AGS cells were seeded in 96-well plates and cultured for 24 h. Cells were treated with B97% and B85% extracts at 10, 20, 50, 100, and 200 μg/mL for 4 and 24 h. After the treatment, the medium was removed, and cells were incubated with fresh medium in the presence of 10% *v/v* CCK-8 (Cell Counting Kit, Merck) [43,44]. The assay is based on the reduction of a water-soluble tetrazolium salt WST-8 operated by cell dehydrogenases that lead to formazan production, soluble and orange-colored; the absorbance of the formazan dye is measured at 450 nm with a microplate reader VICTOR® Nivo™3s (Perkin-Elmer, Waltham, MA, USA). Cell viability in treated groups was compared to untreated cells (control). Two independent tests were performed ($n = 8$).

2.7. Cytotoxicity Assay in Presence of CYP450 Substrates

Hepatic HepG2 cells were seeded in 96-well plates and cultured for 24 h. Firstly, cell viability was tested for the CYP450 substrates, specifically phenacetin (CYP1A2), dextromethorphan (CYP2D6), and triazolam (CYP3A4) at the concentrations of 0.1, 1, 10, and 50 μg/mL for 24 h (all drugs were purchased from Merck). After evaluating non-cytotoxic concentrations, HepG2 cells were seeded in 96-well plates and cultured, then treated with B97% and B85% at 50 or 100 μg/mL in the presence of phenacetin, dextromethorphan, or triazolam at 20 μg/mL for 24 h. Cell viability was assessed by means of the CCK-8 kit as described above.

2.8. Dosage of Intracellular Reactive Oxygen Species (ROS) Level

ROS production was quantified using 2′,7′-dichlorodihydrofluorescein diacetate (H2-DCF-DA, Thermo Fischer Scientific, Waltham, MA, USA) [45]. HEK293 cells were seeded in 96-well plates and cultured for 24 h. Cells were treated with B97% and B85% extracts at 10, 20, and 50 μg/mL for 24 h. H_2O_2, 0.5 mM, was used as a positive control of ROS production. After the treatment, the medium was removed, and cells were washed twice with PBS, then incubated with a 50 μM H2DCF-DA solution for 45 min at 37 °C. In the presence of ROS, the reagent H2DCF-DA was converted in a fluorescent adduct, dichlorofluorescein (DCF). DCF fluorescence intensity was measured at an excitation of 485 nm and emission of 535 nm, using a Multilabel Plate Reader VICTOR® Nivo™3s (Perkin-Elmer). Two independent experiments with four replicates ($n = 8$) were performed.

2.9. Migration Assay

Caco-2 cells (5×10^5) were seeded into 6-well cell culture plates and allowed to grow to 70–80% confluence as a monolayer [43,46,47]. The monolayer was gently scratched across the center of the well with a sterile 1 mL pipette tip. A second scratch was performed perpendicular to the first, creating a cross in each well. After scratching, the medium was removed, and the wells were washed twice in PBS solution. Fresh medium containing 5% *v/v* of heat-inactivated FBS and B97% or B85% at 10 or 100 μg/mL, respectively, were added to each well. Images were obtained from the same fields immediately after scratching (t_0) and after 6, 24, 30, and 48 h using a Leica DMIL microscope, and analyzed using ImageJ software version v1.54j by manually selecting the wound region and recording the total area. The experiments were conducted in triplicate, and two fields were analyzed for each replicate ($n = 4$). Untreated scratched cells represented the control. The percentage of scratching gap was calculated using the following formula: [Wound area t)/Wound area t_0] × 100.

2.10. Total RNA Extraction, Reverse Transcription, and Real-Time PCR

HEK293 cells were seeded in 6-well plates at the density of 2×10^6 cells/well and cultured for 24 h. Cells were stimulated with B97% and B85% at 10 and 20 μg/mL for 24 h. RNA extraction and DNAse treatment were performed as previously described [31,48] by using Ripospin II mini-Kit (#314–150, GeneAll, Seoul, Republic of Korea), according to the manufacturer's protocol. For cDNA synthesis, one microgram of RNA was retrotranscribed with PrimeScript RT Reagent Kit (#RR037A, Takara Bio, Shiga, Japan) and RT-qPCR was performed by CFX Connect Real-Time PCR machine (Bio-Rad Laboratories, Hercules, CA, USA), using SsoAdvanced Universal SYBR Green Supermix (Bio-Rad Laboratories) and specific forward and reverse primers at a final concentration of 300 nM (Table 1). Cycle threshold (Cq) value was determined by the CFX maestro software version 2.3 (Bio-Rad Laboratories), and mRNA expression was calculated by the ΔΔCt method and normalized to the mean of rps18-rpl13a genes as an endogenous control. For gene expression analysis, endogenous control mRNA levels were not affected among treatments ($p > 0.05$, one-way analysis of variance: ANOVA).

Table 1. Transcript and sequence of each primer used in real-time PCR.

Gene	NCBI GenBank	Sequence
Rplp13a	NM_012423.4	fw: GTGCGTCTGAAGCCTACAAG rv: CGTTCTTCTCGGCCTGTTTC
Rps18	NM_022551.3	fw: TCTAGTGATCCCTGAAAAGT rv: AACACCACATGAGCATATC
Tp53	NM_000546.6	fw: AGGGATGTTTGGGAGATGTAAG rv: CCTGGTTAGTACGGTGAAGTG
c-Myc	NM_001354870.1	fw: AAGCTGAGGCACACAAAGA rv: GCTTGGACAGGTTAGGAGTAAA
n-Myc	NM_005378.6	fw: TCCAGCAGATGCCACATAAG rv: ACCTCTCATTACCCAGGATGTA
Met	NM_001127500.3	fw: CCTGGGCACCGAAAGATAAA rv: CTCCTCTGCACCAAGGTAAAC
Mdm2	NM_002392.6	fw: AGGCTGATCTTGAACTCCTAAAC rv: CAGGTGCCTCACATCTGTAATC
Cdkn1a	NM_000389.5	fw: CGGAACAAGGAGTCAGACATT rv: AGTGCCAGGAAAGACAACTAC
Snai1	NM_005985.4	fw: CAGATGAGGACAGTGGGAAAG rv: GAGACTGAAGTAGAGGAGAAGGA
Snai2	NM_003068.5	fw: AACTACAGCGAACTGGACAC rv: GAGGATCTCTGGTTGTGGTATG
Hras	NM_005343.4	fw: AAGCAAGGAAGGAAGGAAGG rv: GTGGCATTTGGGATGTTCAAG
Cdk4	NM_000075.4	fw: GCTCTGCAGCACTCTTATCTAC rv: CTCAGTGTCCAGAAGGGAAATG
Bax	NM_004324	fw: CTCCCCATCTTCAGATCATCAG rv: GGCAGAAGGCACTAATCAAGTC
Bcl2	NM_000657	fw: GACTGAGTACCTGAACCGGC rv: CTCAGCCCAGACTCACATCA

2.11. Statistical Analysis

Data were presented as mean ± standard deviation (SD). Statistical analyses were performed using the unpaired Student *t*-test or one-way analysis of variance (ANOVA) (with $p < 0.05$ significance level) as appropriate, followed by Dunnet post-hoc tests for multiple comparison. Analyses and graphs were conducted and composed by using GraphPad Prism 10.1 (San Diego, CA, USA).

3. Results

3.1. Chemical Analyses of Extracts Containing Berberine and Protoberberinoids

Chemical analysis of sixteen different marketed herbal preparations containing BER and PROTBERs were performed, specifically extracts from *B. aristata*, namely *B. aristata* bark extracts with a declared content of BER of 97% *w*/*w* db, and *B. aristata* bark extracts with a declared content of BER of 85% *w*/*w* db. A new method developed on purpose was used, providing a robust and reliable performance for the extracts under investigation containing BER and PROTBERs (Table 2).

Table 2. Parameters of HPLC-DAD method employed.

Parameter	Value
R^2	0.99
Equation	y = 4317.80 − 98.57
Linearity range	0.04–7.5 µg in column
Recovery of spiked standard	>90% and <105%
Limit of quantification	0.03 µg in column
Intra- and inter-day variation	<3%

The method allowed us to perfectly identify and quantify BER (as BER hydrochloride) in all samples analyzed at the retention time (RT) of 5.7 min (Table 3).

Table 3. Quantification of berberine in 16 different marketed herbal preparations containing berberine and protoberberinoids by means of HPLC-DAD. Quantification was made on dry basis (db) of the extracts, and values are expressed as berberine hydrochloride. Values are expressed as % *w*/*w* of a triplicate analysis performed on eight *B. aristata* extracts with a declared content of 97% of berberine hydrochloride db and on eight extracts of *B. aristata* with a declared content of 85% of berberine hydrochloride.

Sample	Berberine % *w/w* db	Sample	Berberine % *w/w* db
A85	86.26	A97	97.41
B85	86.74	B97	91.94
C85	91.93	C97	96.50
D85	88.87	D97	97.16
E85	89.63	E97	97.49
F85	90.66	F97	97.21
G85	80.07	G97	97.29
J85	87.88	J97	97.80
mean ± SD	87.76 ± 3.64	mean ± SD	96.60 ± 1.91

We also identified jatrorrhizine in all the samples (RT = 5.0 min), as well as palmatine (RT = 5.3 min), in agreement with previously published works [49] and with the characteristic UV–vis spectra. A partial overlapping of palmatine with another minor PROTBERs was noted, and thanks to the collaboration of external labs and by using a mass spectrometer, we identified this PROTBER as berberrubine. In many samples, we found other secondary PROTBERs through their maximum absorbance at 344–347 nm, but the sum of these compounds expressed as BER hydrochloride was confirmed to be under the limit of quantification (LOQ) (Figure 2).

Summarizing, in all samples, we found BER as a major compound, followed by jatrorrhizine, always present, and palmatine at low concentrations, often accompanied by berberrubine (Table 4). The overall qualitative and quantitative data proved the extracts were true to the label for both the 97% and 85% groups, with a maximum discordance of +8.2% for C85 and −5.8% for G85. The higher variability in BER content in samples labeled with BER 85% compared to those labeled with BER 97% was noted; the reason for this are not known to date, but we could generally attribute it to differences in the method of extract preparation.

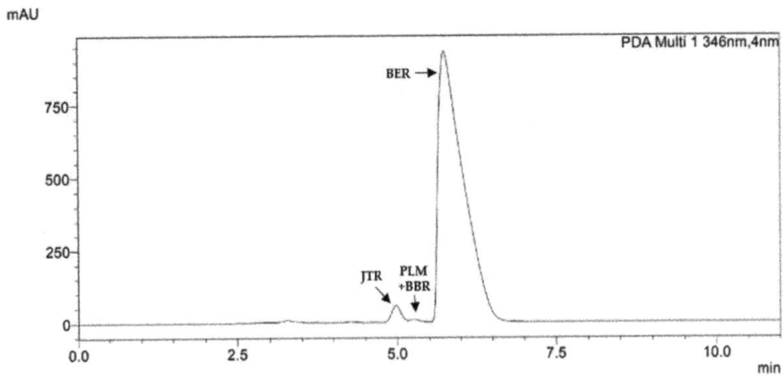

Figure 2. HPLC chromatogram of a representative *Berberis aristata* DC. bark extract recorded at 346 nm. The main peak (retention time 5.7 min) is related to berberine (BER); at 5.0 min, there is jatrorrhizine (JTR). Among these two molecules, at 5.3 min, two partially overlapped peaks are present, attributed to berberrubine (BBR) and palmatine (PLM).

Table 4. Quantification of protoberberinoids in *B. aristata* bark extracts. Samples are labeled according to the declared content of berberine expressed as berberine hydrochloride. Quantification was made on dry basis (db) of the extracts and all compounds are expressed as berberine hydrochloride. Values are expressed as % *w*/*w* of a triplicate analysis performed on eight *B. aristata* extracts with a declared content of 97% of berberine hydrochloride db and on eight extracts of *B. aristata* with a declared content of 85% of berberine hydrochloride. The quantification of palmatine and berberrubine was performed by repeating chromatographic runs of high amounts of injected samples in order to reach the limit of quantification.

Samples	Jatrorrhizine % *w/w* db	Berberrubine + Palmatine % *w/w* db
A85	2.39	0.25
B85	1.88	0.24
C85	2.10	0.23
D85	3.07	0.26
E85	2.54	0.28
F85	2.85	0.24
G85	2.59	0.20
J85	3.12	0.25
mean ± SD	2.57 ± 0.44	0.24 ± 0.02
A97	2.25	0.24
B97	1.49	0.08
C97	2.98	0.11
D97	2.57	0.23
E97	2.04	0.13
F97	2.04	0.14
G97	2.05	0.28
J97	1.71	0.13
mean ± SD	2.14 ± 0.46	0.17 ± 0.07

Contrarily to what emerged for other botanicals, the quality of *B. aristata* extracts available in the raw ingredients marketplace is reliable [50]. This evidence, however, is seemingly in contrast with recent investigations in which only 56% of 18 *Berberis*-derived food supplements were true-to-the-label [27]. It must be specified in this regard that our samples were crude extracts available in the business-to-business ingredients market rather than retail, formulated food supplements.

Given the homogeneous phytochemical profile, for subsequent tests, two *B. aristata* extracts with a declared content of 97% of berberine hydrochloride db and two extracts of *B. aristata* with a declared content of 85% of berberine hydrochloride were chosen randomly (A85, D85, A97, G97) and pooled to obtain the samples B85 and B97, whose chemical composition is reported in Table 5.

Table 5. Chemical composition of pooled A85 and D85 (mixed 1:1, B85) and pooled A97 and G97 (1:1, B97) expressed as mean of composition of single extracts.

Samples	Berberine % *w/w* db	Jatrorrhizine % *w/w* db	Berberrubine + Palmatine % *w/w* db
B. aristata 85% (B85)	87.57 ± 1.85	2.83 ± 0.48	0.26 ± 0.01
B. aristata 97% (B97)	97.35 ± 0.08	2.15 ± 0.14	0.26 ± 0.02

3.2. In Vitro Bioaccessibility Assessment of Berberine and Protoberberinoids

The assessment of simulated digestion on the different *B. aristata* extracts examined allowed us to recover BER and jatrorrhizine in all samples (Table 6). The post-digestive presence of bile salts and enzymes partly hid the minority part of berberrubine and palmatine in some replicates; thus, we decided not to report these PROTBERs in the assessment of bioaccessibility. We could verify that isoquinolinic alkaloids underwent minimal degradation; indeed, the post-digestive bioaccessibility rate was >95% for BER and 84–88% for jatrorrhizine, with low differences in different samples (Table 6). BER was adequately soluble in hydrophilic gastrointestinal digestive fluids, and it was not affected by the activity of digestive enzymes. Figure 3A,B shows the stability obtained from the analysis of a representative extract before and after simulated digestion.

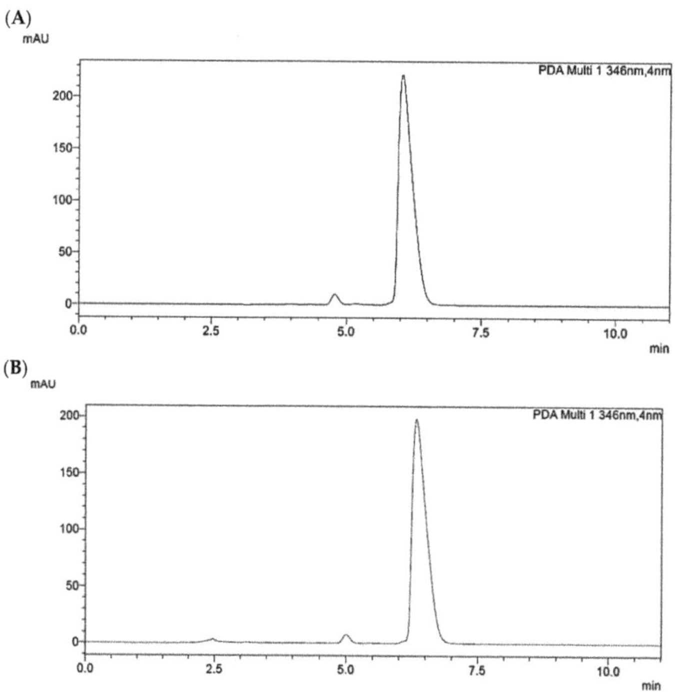

Figure 3. Representative HPLC chromatogram of *Berberis aristata* DC. bark extract containing 85% of berberine hydrochloride dry basis before (**A**) and after (**B**) simulated digestion, recorded at 346 nm.

Table 6. Bioaccessibility rate of berberine and jatrorrhizine investigated in two extracts of *Berberis aristata* DC. bark containing 97% or 85% of berberine hydrochloride dry basis, respectively. Berberine showed high stability in response to pH change and digestive enzyme activity; jatrorrhizine was also recovered in high percentage.

Samples	Bioaccessibility Rate %	
	Berberine	**Jatrorrhizine**
B. aristata 97%	>95	83.57 ± 3.33
B. aristata 85%	>95	87.81 ± 3.69

3.3. In Silico Pharmacokinetic Analysis and Target Prediction of Berberine and Protoberberine Derivatives

After testing the bioaccessibility of BER and PROTBERs, we intended to study other pharmacokinetic features of BER, jatrorrhizine, berberrubine, and palmatine that are stable and present at detectable levels in the most used herbal preparations containing BER and PROTBERs.

By means of SwissADME® tools, we verified that BER (both in neutral and in hydrochloride forms) has good potential to be absorbed by intestinal epithelium, but being a P-gp substrate explains data referring to its poor bioavailability [30,51]. Moreover, the computational prediction confirmed that BER is a substrate for the following CYP450 isoforms: 1A2, 3A4, and 2D6 (Figure 4).

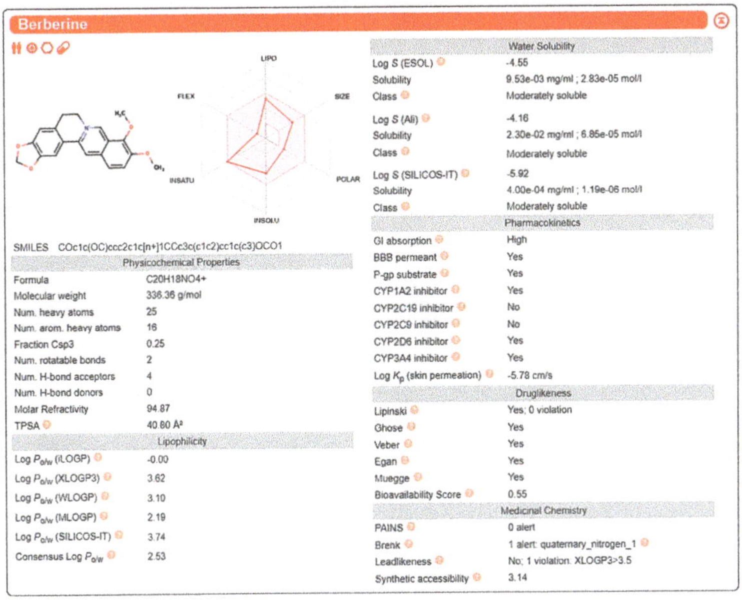

Figure 4. Computational analysis of berberine pharmacokinetic properties by means of SwissADME tool (http://www.swissadme.ch/, accessed on 20 January 2024).

Jatrorrhizine and berberrubine were shown to share almost all pharmacokinetic characteristics with BER (Supplementary Figure S1A,B). As observed for BER, jatrorrhizine, berberrubine, and palmatine resulted in a P-gp, CYP2D6, and CYP3A4 substrate, but not CYP1A2; moreover, palmatine was the least soluble PROTBER among those examined (Supplementary Figure S1C).

We used the GeneCards tool to collect known targets for BER and the other PROTBERs investigated here, and we found a total of 371 targets for BER, 6 for jatrorrhizine, 11 for berberrubine, and 16 for palmatine. Interestingly, the overlapping of targets found for the

4 compounds produced 373 results, thereby demonstrating that almost all targets known for BER are shared with other major PROTBERs of *Berberis* spp. (Figure 5).

	Symbol	Description	Category	UniProt ID	GIFtS	GC Id	Score
1	LDLR	Low Density Lipoprotein Receptor	Protein Coding	P01130	55	GC19P095238	9.12
2	BCL2L1	BCL2 Like 1	Protein Coding	Q07817	52	GC20M031664	6.86
3	CCND1	Cyclin D1	Protein Coding	P24385	57	GC11P069641	6.83
4	LINC01672	Long Intergenic Non-Protein Coding RNA 1672	RNA Gene		13	GC01P008900	6.81
5	SLC2A4	Solute Carrier Family 2 Member 4	Protein Coding	P14672	50	GC17P092348	6.69
6	CDK4	Cyclin Dependent Kinase 4	Protein Coding	P11802	57	GC12M058990	6.40
7	CASP8	Caspase 8	Protein Coding	Q14790	55	GC02P201233	6.40
8	WEE1	WEE1 G2 Checkpoint Kinase	Protein Coding	P30291	52	GC11P009573	5.81
9	CDK1	Cyclin Dependent Kinase 1	Protein Coding	P06493	51	GC10P060772	5.81
10	BDNF-AS	BDNF Antisense RNA	RNA Gene		22	GC11P027466	5.55
11	CYP2D6	Cytochrome P450 Family 2 Subfamily D Member 6	Protein Coding	P10635	50	GC22M042126	5.29
12	CDK6	Cyclin Dependent Kinase 6	Protein Coding	Q00534	54	GC07M092604	5.23
13	CDK2	Cyclin Dependent Kinase 2	Protein Coding	P24941	54	GC12P055966	5.20
14	CCNB1	Cyclin B1	Protein Coding	P14635	51	GC05P069167	5.20
15	BIRC5	Baculoviral IAP Repeat Containing 5	Protein Coding	O15392	50	GC17P078214	5.16
16	NFE2L2	NFE2 Like BZIP Transcription Factor 2	Protein Coding	Q16236	54	GC02M177227	5.12
17	LPL	Lipoprotein Lipase	Protein Coding	P06858	53	GC08P019901	4.90
18	MIR7-3HG	MIR7-3 Host Gene	RNA Gene	Q8N6C7	27	GC19P094936	4.75
19	TMX2-CTNND1	TMX2-CTNND1 Readthrough (NMD Candidate)	RNA Gene		17	GC11P057712	4.53
20	CASC2	Cancer Susceptibility 2	RNA Gene	Q8IU53	26	GC10P118053	4.52

Figure 5. The best score for analysis of target prediction for berberine, as predicted by GeneCards tool (https://www.genecards.org/, accessed on 23 January 2024). Predicted targets are very heterogeneous but mainly refer to the most studied biological activities of berberine.

We observed that the main targets identified by GeneCards were related to low-density lipoprotein receptor (LDLR) and antiproliferative activity, for which BER activity is known [52–55]. Other targets were related to CYP450 interaction (CYP2D6) and anti-neuroinflammatory activity (BDNF, NFE2), as already reported in pre-clinical studies [56] (Figure 5).

The interaction between BER and jatrorrhizine mainly underlined the activity towards neuroprotective-related targets (BDNF and acetylcholinesterase, ACHE) and on cell cycle regulatory factors (Supplementary Figure S2A).

Differently, the analysis of the interaction between BER and berberrubine highlighted the anti-inflammatory activity of berberrubine [57] (Supplementary Figure S2B).

As regards the combination BER with palmatine, we found that for some newly emerged targets, such as those related to cell cycle regulation and antiproliferative activity, corroborative data emerged from an In Vitro study on cancer cells [58–61]; moreover, BDNF and cholinesterase (BCHE) emerged as neuro-targets, and some important antioxidant targets such as catalase (CAT) and superoxide dismutase (SOD) were shared by BER and palmatine, as well as the xenobiotic toxicity modulator aryl-hydrocarbon receptor (ARH) [62]. CYP1A(1-2) resulted to be modulated by BER and palmatine (Supplementary Figure S2C).

Swiss and SEA target prediction, set as free search, without organ or signaling restrictions, confirmed that BER has a strong probability to interact with CYP2D6, as already described, but also with CYP1B1 and CYP1A2 (predicted by STP and SEA, respectively) (Supplementary Figure S3A). Both prediction tools also indicated acetylcholineesterase and cholinesterases (ACHE, BCHE) as targets, and the interaction with Ras-related C3 botulinum toxin substrate 1 (RAC1), a member of Rho GTPase (Supplementary Figure S3A). These predictions, only in part already known, may explain why BER is currently considered in the field of neurodegenerative disorders [63], vascular system [64], and metabolism regulation [24].

Regarding jatrorrhizine, the tools predicted affinity with medium or low scores, but ACHE was the most plausible target for this PROTBER. STP and SEA shared the prediction of RAC1, involved in metabolism, and cell division control protein 42 homolog (CDC42), a regulator of the cell cycle (Supplementary Figure S3B).

Prediction scores for berberrubine were the worst, being weak only for humans; STP predicted with a medium or low score some targets already considered for BER, such as RAC1 and CDC42 (Supplementary Figure S3C).

As observed for BER, palmatine interacted with ACHE, as the data experimentally confirmed [65]. Other targets such as 5-HTRB2, BCHE, ADRA2, CHRM1, SIGMAR1, and CYP2D6 were predicted with a medium score by SwissTarget, and RAC1 by SEA with a low score (Supplementary Figure S3D).

3.4. *In Vitro Cytotoxicity Evaluation of Berberis aristata Bark Extracts*

To assess the biological safety of BER and PROTBER alkaloids, we investigated the biological impact of both extracts (B97% and B85%) examined so far. Specifically, we evaluated the possible cytotoxic effects of BER- and PROTBERs-containing preparations by performing a cell viability assay on an In Vitro model of different human cell lines: intestinal (Caco-2), hepatic (HepG2), gastric (AGS), and renal (HEK293). Cells were treated with B97% and B85% at different concentrations (10, 20, 50, 100, and 200 μg/mL) at two time points, a short- (4 h) and a long-term (24 h), to simulate the exposure time of different organism systems (stomach, intestine, liver, and kidney) and cell types to BER and PROTBERs. The results obtained by the cell viability assay allowed us to determine the IC_{50} (half-maximal inhibitory concentration) value for each cell line at both time points of treatment. After 4 h of treatment, in all cell lines, the IC_{50} values were over 100 μg/mL; specifically, for the hepatic (HepG2) and gastric (AGS) cells, the IC_{50} values exceeded 200 μg/mL (Table 7). Similarly, after 24 h of treatment, the IC_{50} values were still higher than 100 μg/mL despite the increasing time of treatment. Of note, differences in the extracts tested were very low, demonstrating a similar cell impact of the different *B. aristata* extracts studied here. Differences in cell viability were observed in the various cell line models, but they could be considered non-significant given the high IC_{50} value obtained (Table 7). These data suggested a negligible cytotoxic impact of BER-containing herbal preparations, providing important outcomes on the safety of the use of BER and PROTBERs contained in the most common marketed food supplements.

Table 7. IC50 values calculated for the extracts of *B. aristata* bark containing berberine hydrochloride 97% (B97) and 85% (B85) by means cell viability assay performed on intestinal (Caco-2), gastric (AGS), hepatic (HepG2), and kidney (HEK293) cells after 4 h and 24 h of treatment.

		IC_{50} (μg/mL)			
Sample	**Treatment (h)**	**AGS**	**Caco-2**	**HepG2**	**HEK293**
B97%	4	>200	166.93 ± 8.44	>200	142.65 ± 12.01
B85%	4	>200	169.14 ± 9.25	>200	181.93 ± 24.50
B97%	24	>200	105.59 ± 11.21	198.85 ± 8.50	127.16 ± 25.18
B85%	24	>200	107.34 ± 9.68	186.41 ± 8.42	143.43 ± 29.80

Data are expressed in μg/mL as mean ± standard deviation (SD).

3.5. *In Vitro Cytotoxicity Evaluation of Berberis aristata Bark Extracts in Presence of CYP450 Substrates*

Considering the hepatic metabolism of BER and PROTBERs mediated by CYP450, our investigation included the possible toxicological interactions with other drugs known to be substrates of CYP450 isoforms.

To address this aspect, we performed a cell viability assay on HepG2 cells treated with both B97% and B85% extracts at the highest concentrations of 50 and 100 μg/mL in the presence of three main CYP450 substrates, like phenacetin (P), dextromethorphan (D), and triazolam (T).

The selection of concentrations derived from the cytotoxicity tests described in 3.4 were intended to create a stress condition that could better assess the possible effects of interaction with other drugs at the cytotoxic level. Firstly, to identify the non-toxic concentration of P, D, and T to use in the co-treatment, we tested HepG2 cell viability treated with different concentrations of P, D, and T (0.1, 1, 10, and 50 μg/mL) for 24 h. Statistical analysis revealed that all three drugs showed no impact on cell viability up to 10 μg/mL compared to the control group (Figure 6A–C). At the dose of 50 μg/mL, P, D, and T differently affected the viability of hepatic cells: P and D reduced cell viability by about 10% and 20%, respectively (one-way ANOVA: P = $p < 0.05$; D = $p < 0.001$ vs. CTRL), while T showed the most significant impact by reducing cell viability of about 90% with respect to controls ($p < 0.0001$ vs. CTRL) (Figure 6C). Based on these data, we carried out following analysis by treating hepatic cells with 20 μg/mL of P, D or T in association with B97% or B85% at 50 or 100 μg/mL, respectively. The results showed a non-significant effect of B97% at 50 μg/mL on HepG2 cell viability; the association with P and T at 20 μg/mL did not cause changes in cell viability when compared to the control group, while the effect of BER was slightly worsened by the presence of D, resulting in a reduction in cell viability of up to 30% (one way ANOVA: $p < 0.0001$ vs. CTRL) (Figure 6D). The treatment with B97% at 100 μg/mL significantly affected cell viability compared to controls ($p < 0.001$ vs. CTRL), but it was not significantly altered by the co-presence of the three drugs (Figure 6D). Regarding B85%, the exposure of HepG2 cells to the sample alone slightly reduced the cell viability (−17% compared to control) that did not undergo alterations in the presence of P, D, or T when compared to the control cells (Figure 6E). The 24 h treatment of hepatic cells with B85% at 100 μg/mL significantly reduced the cell viability by about 30% with respect to the controls ($p < 0.0001$ vs. CTRL). Also, in this case, the association with all drugs did not change the effect of BER (Figure 6E); indeed, the co-presence of P seemed to improve the impact of B85% on cell viability ($p < 0.05$ vs. B85% at 100 μg/mL).

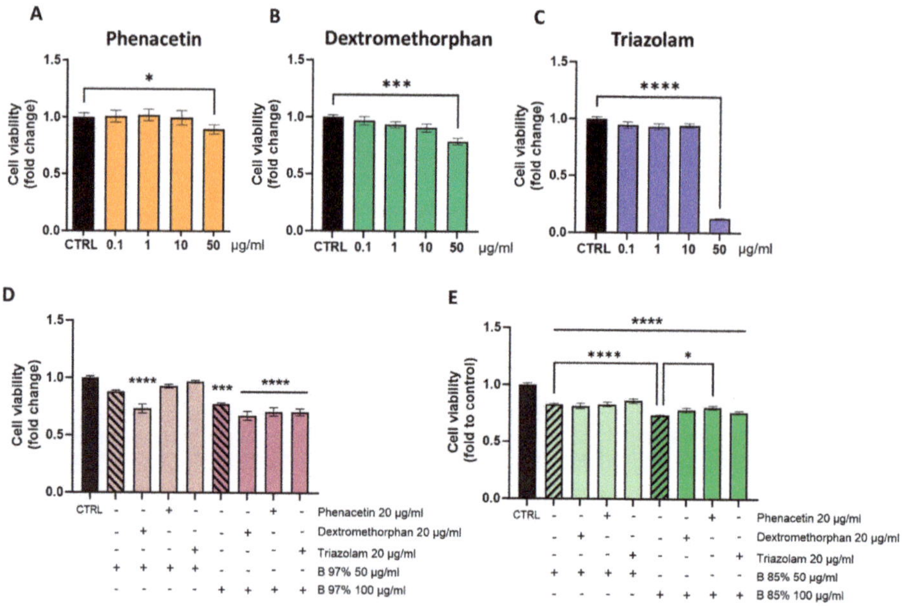

Figure 6. Cell viability assay on HepG2 cells after 24 h of treatment with different concentrations of (**A**) phenacetin, (**B**) dextromethorphan, and (**C**) triazolam. Then, HepG2 cells were co-treated for 24 h with phenacetin, dextromethorphan, or triazolam at 20 μg/mL and B97% at 50 and 100 μg/mL (**D**) or B85% at 50 μg/mL (light colour) and 100 μg/mL (dark colour)(**E**), respectively. The column with the striped pattern represents the treatment with only the extract. Each column represents mean ± SEM. Data were analyzed by one-way analysis of variance followed by Dunnet post-hoc: *** $p < 0.001$ and **** $p < 0.0001$ vs. CTRL; **** $p < 0.0001$ vs. B85% 50 μg/mL; * $p < 0.05$ vs. B85% 100 μg/mL; $n = 4$.

These results showed that, in association with P, D, or T, the low cytotoxicity of the most common BER-containing extracts remained unchanged. Overall, these findings provided preliminary data on toxicological aspects related to the interaction of B97% and B85% extracts with drugs known to be CYP450 substrates.

3.6. Effect of Berberis aristata Bark Extracts on Cell Migration

In addition to assessing the effects on cell viability, the investigation of the safety of herbal preparations containing BER and PROTBERs proceeded by analyzing cell migratory movement by means of the wound-healing assay. This is a commonly used method to determine the effects of compounds on changes in 2D cell migration properties, relating in particular to cancer cells.

With this purpose, we explored the migratory activity of human colorectal cancer cells (Caco-2) following treatment with B97% and B85% at both low concentration (10 µg/mL) and high concentration (100 µg/mL) for 48 h. The relative scratch gap was monitored over time and measured. Data showed a statistically significant effect of both B97% and B85% at the dose of 100 µg/mL in decreasing the migration ability of Caco-2 cells (about 40% B97% and 33% B85%) at the end of the treatment compared to control cells (one-way ANOVA: ** $p < 0.01$ vs. CTRL) (Figure 7). These results suggested a functional activity of tested extracts at 100 µg/mL in decreasing the cell migration of intestinal tumor cells.

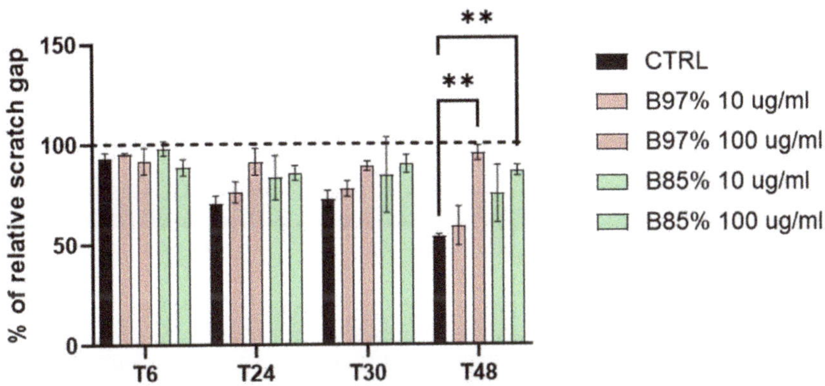

Figure 7. Wound-healing assay on Caco-2 cells treated with B97% and B85% (10 and 100 µg/mL) for 48 h or untreated (CTRL). Dashed line represent the 100% of wound area. Each column represents mean ± SEM. Data were analyzed by one-way analysis of variance: ** $p < 0.01$ vs. CTRL; $n = 4$.

3.7. In Vitro Evaluation of ROS Production in Normal Kidney Cells Treated with Berberis aristata Bark Extracts

The assessment of the safety of herbal preparations containing BER and PROTBERs also took into consideration the possible effects of oxidative stress of both B97% and B85% extracts in non-tumoral human cells. For this reason, ROS production was evaluated in normal renal cells (HEK293) treated for 24 h with B97% and B85% extracts at different concentrations (10, 20, and 50 µg/mL); H_2O_2 was used as a positive control.

The results showed no statistical effects of either B97% or B85% on any of the concentrations tested compared to the control cells; in contrast, treatment with H_2O_2 was able to induce ROS production by 1.45-fold with respect to untreated cells (one-way ANOVA: $p < 0.0001$ vs. CTRL) (Figure 8).

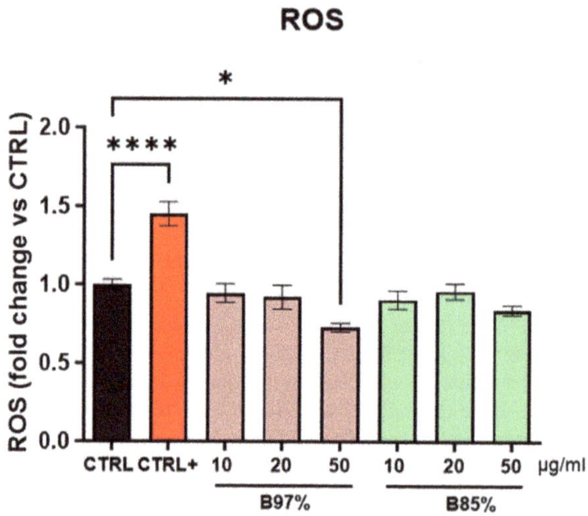

Figure 8. Dosage of intracellular ROS in HEK293 cells treated with B97% and B85% at the concentrations of 10, 20, and 50 µg/mL for 24 h. H_2O_2 (500 µM) was used as positive control (CTRL+). Each column represents mean ± SEM. Data were analyzed by one-way analysis of variance: * $p < 0.05$ and **** $p < 0.0001$ vs CTRL; $n = 8$.

These data revealed no effects of BER or PROTBERs in inducing oxidative stress through ROS production.

3.8. Transcriptional Effects of *Berberis aristata* Bark Extracts on Target Genes Involved in Cell Cycle Control and Neoplastic Transformation

With the purpose of deepening the study on the safety and biological activity of BER, we investigated the molecular effects of BER- and PROTBERs-containing herbal preparations in an In Vitro model of human normal kidney cells, HEK293.

We focused on the transcriptional effects on the main target genes involved in the control of cell cycle, cell growth, neoplastic transformation, and oncogenes by exposing HEK293 cells for 24 h to the two examined extracts, B97% and B85%, at the concentrations of 10 and 20 µg/mL. First, we analyzed through RT-qPCR the gene expression of the tumor protein p53 (*TP53*), the oncogene *MDM2*, and the proto-oncogenes *c-MYC*, *n-MYC*, *HRAS*, and *MET*.

Our results showed that *TP53*, *MDM2*, and *n-MYC* mRNA levels were markedly downregulated in cells treated with B97% and B85% at both 10 and 20 µg/mL compared to the control counterpart [one-way ANOVA: *TP53*: F (4, 18) = 12.04, $p < 0.001$; *MDM2*: F (4, 18) = 7.773, $p = 0.0008$; *n-MYC:* F (4, 18) = 8.198, $p = 0.0006$] (Figure 9A–C). On the contrary, no differences were observed in *c-MYC* gene expression with respect to untreated cells with the exception of cells exposed to B97% at 20 µg/mL where a significant increase (2-fold) in *c-MYC* mRNA levels, compared to the control, was observed [F (4, 18) = 7.375; $p = 0.0011$] (Figure 9D). Regarding the mRNA levels of *HRAS* and *MET*, we found no significant effects of either BER extracts, at either concentration, in kidney cells following 24 h of treatment (Figure 9E,F). These results revealed the impact of both *B. aristata* extracts in downregulating the gene transcription of *TP53*, *MDM2*, and *n-MYC* at both the concentrations tested in normal cells.

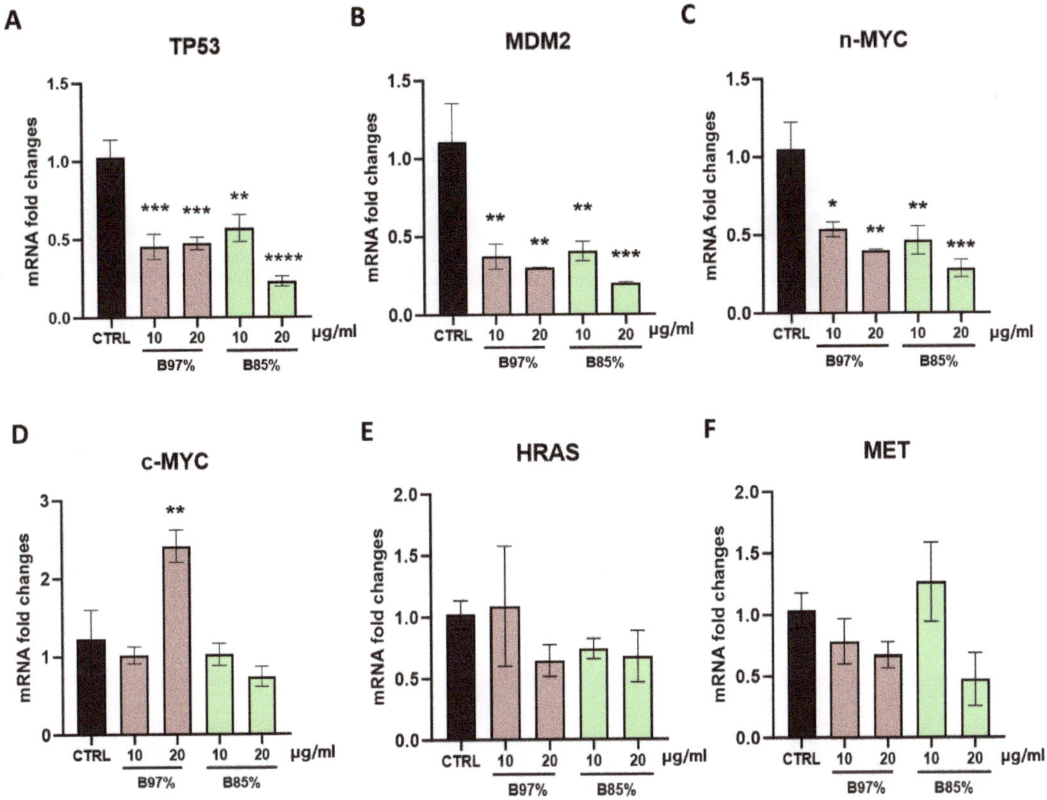

Figure 9. RT-qPCR analysis of *TP53* (**A**), *MDM2* (**B**), *n-MYC* (**C**), *c-MYC* (**D**), *HRAS* (**E**), and *MET* (**F**) transcripts in HEK293 cells following treatment with B97% and B85% at 10 or 20 μg/mL. Each column represents mean ± SEM. Data were analyzed by one-way analysis of variance followed by Dunnet post-hoc: * $p < 0.05$, ** $p < 0.01$, *** $p < 0.001$, and **** $p < 0.0001$ vs. CTRL ($n = 4$).

Then, our molecular analysis focused on p21 (*CDKN1A*) and *CDK4* targets, two essential factors in regulating cell cycle. The results revealed no statistically significant effects of the B97% extract on modulating *CDKN1A* gene transcription that was downregulated instead by B85% at both concentrations [F (4, 18) = 4.273, $p = 0.0132$]. Regarding *CDK4*, treatment with both BER extracts induced a decrease in gene expression unrelated to the dose [F (4, 18) = 3.036, $p = 0.0446$] (Figure 10A,B).

In evaluating the effects of BER in regulating cell death processes, such as apoptosis, we also analyzed the transcriptional levels of the Bcl-2 family members, key factors in the early stages of apoptotic process: the pro-apoptotic gene *BAX* and the anti-apoptotic gene *BCL-2* [66–68].

Statistical analysis showed a significant and similar effect of both *B. aristata* extracts in downregulating the expression of the pro-apoptotic gene *BAX* in HEK293 cells compared to controls [one-way ANOVA: F (4, 9) = 8.688, $p = 0.0037$], whereas no changes in *BCL-2* mRNA levels were observed (Figure 11A,B).

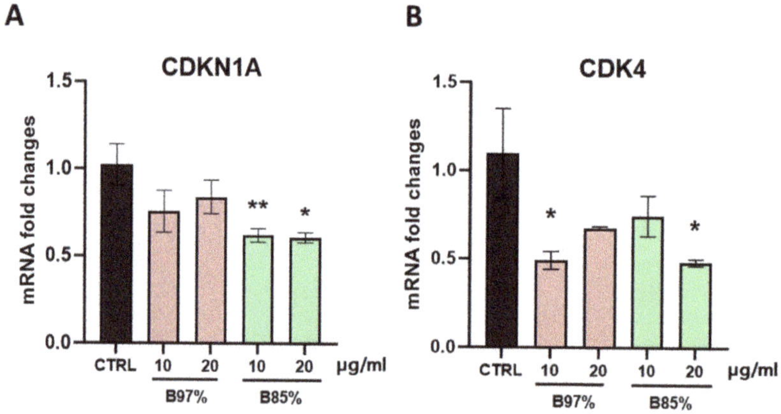

Figure 10. RT-qPCR analysis of *CDKN1A* (**A**) and *CDK4* (**B**) transcripts in HEK293 cells following treatment with B97% and B85% at 10 or 20 µg/mL. Each column represents mean ± SEM. Data were analyzed by one-way analysis of variance followed by Dunnet post-hoc: * $p < 0.05$ and ** $p < 0.01$ vs. CTRL ($n = 4$).

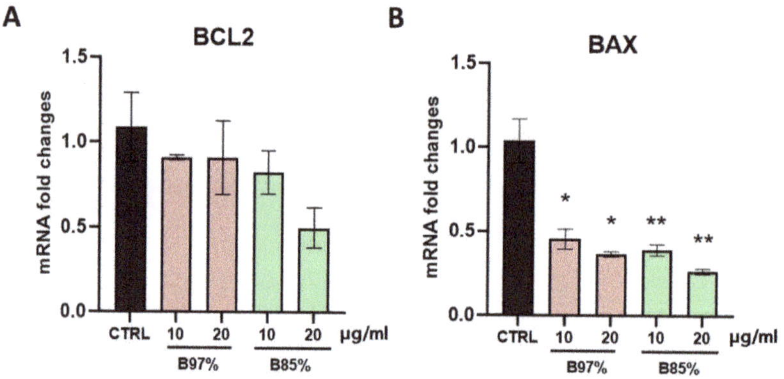

Figure 11. RT-qPCR analysis of *BAX* (**A**) and *BCL-2* (**B**) transcripts in HEK293 cells following treatment with B97% and B85% at 10 or 20 µg/mL. Each column represents mean ± SEM. Data were analyzed by one-way analysis of variance followed by Dunnet post-hoc: * $p < 0.05$ and ** $p < 0.01$ vs. CTRL ($n = 4$).

Accumulating evidence has revealed that BER exerts potential pro-apoptotic effects by modulating *BAX/BCL2* (pro-/anti-apoptotic) expression in multiple cancers, including breast, lung, liver, gastric, colorectal, pancreatic, and ovarian cancers [12,60,69–74]. Our results in normal cells counteracted this evidence and highlighted a *BAX* transcriptional inhibition.

Overall, our data showed that both *B. aristata* extracts examined did not affect or downregulated the transcription of several gene targets involved in the regulation of cell cycle or with oncogenic functions, suggesting an inhibitory activity of BER in cell cycle, proliferation, and division processes in normal cells.

4. Discussion

Herbal preparations containing berberine (BER) and other minor protoberberinoids (PROTBERs) have been extensively used in traditional Asian medicine, in modern conventional phytotherapy, and in food supplements supporting physiological activities. BER has garnered attention for its purported effects on glucose and lipid metabolism, gastrointestinal health, and even for potential anticancer properties [11,15–17,75,76]. However, as the popularity of herbal preparations containing BER and PROTBERs increase, concerns about their biological safety are emerging, especially regarding the safe exposure for general consumers and special population, as highlighted by some national authorities and EFSA in the EU [37]. The authorities have clearly claimed the need for analytical, pharmacokinetic, and biological insights into herbal preparations containing BER and PROTBERs used as food supplements. This study aimed to investigate these important concerns to provide valuable knowledge to the authorities, the scientific community, and companies producing BER-based preparations.

The chemical characterization of *B. aristata* crude extracts available in the market (the most used preparations containing BER and PROTBERs used in food supplements) allowed to confirm the declared content of BER and the identification of jatrorrhizine as the main secondary PROTBER, and palmatine and berberrubine as other minority PROTBERs present in the extracts.

These findings showed that herbal preparations containing BER and PROTBERs are more homogeneous than expected within the EU, being basically referred to as *B. aristata* dry extracts; moreover, BER represents more than 90% of total alkaloids in the most common extracts available in the market.

Pivotal evidence about the pharmacokinetic properties of BER have emerged from In Vitro assays and In Silico predictions; BER and PROTBERs are only minimally degraded before intestinal absorption, but being a P-gp substrate and metabolized by cytochrome P450 (CYP450), with particular regard to the CYP2D6, CYP1A2, and CYP3A4 isoforms, a low bioavailability is expected in particular for BER [35,52,61]. Computational predictions were confirmed experimentally for BER, and in part for palmatine, whereas a mainly structural-homology-based prediction could be obtained for jatrorrhizine and berberrubine. Besides LDLR, well known to be peculiar target of BER [77], all computational tools confirmed the activity of BER and PROTBERs towards neuroprotective [55,62,64,78,79] and antiproliferative targets [59–61,80,81].

These data allowed us to move towards a rational and deeper In Vitro safety assessment of the plausible concentrations of BER and PROTBERs that are able to reach different organism districts after oral administration. Noticeably, the target prediction and cell and molecular analysis addressed in this study were focused on human models, recognizing the limitations and misleading interpretations in extrapolating findings from animal studies to human safety [71]. Therefore, given the homogenous phytochemical profile available, we evaluated the effects of two representative *B. aristata* bark dry extracts on gastric (AGS), intestinal (Caco-2), hepatic (HepG2), and renal (HEK293) cells, following short- or a long-term exposure (4 and 24 h). The results for both exposures showed a very limited impact of BER-enriched extracts on cell viability, with an IC_{50} always exceeding 100 μg/mL in all cell lines examined. We considered cell viability only as a preliminary but fundamental test to assess the safe concentrations of samples under investigation, allowing us to quickly move towards the investigation of the interaction effects between BER and PROTBERs and CYP1A2 -2D6 and -3A4 substrates on hepatic cell viability. Interestingly, even at 50 and 100 μg/mL, the *B. aristata* bark dry extracts did not affect cell viability in the presence of phenacetin (CYP1A2), dextromethorphan (CYP2D6), or triazolam (CYP3A4), demonstrating that herbal preparations containing BER and PROTBERs may not have negative interactions with CYP450 substrates.

Many In Vitro studies demonstrated the ability of BER to inhibit cell proliferation, invasion, and to regulate the cell cycle in tumoral cells [59,61,80–83]. In light of this, our study proceeded with the analysis of cell response to the exposure to both BER extracts by evaluating the cell migration of colorectal adenocarcinoma cells (Caco-2) through the wound-healing assay. This test is a commonly used method in cell biology to determine the effects of compounds on the changes in migration properties, an important feature of cells to move and invade other tissues, interact with other cells, or to repair any damage. This property is typical of cancer cells and identifies their invasive and aggressive potential. Our data demonstrated a delay in wound closure when treated with *B. aristata* bark dry extracts, even after a long exposure time, both at 10 and 100 μg/mL, corroborating numerous evidence regarding the BER and PROTBER inhibitory effects of cell migration [80,81,84–86]. Our experimental evidence supported the potential antitumoral activity of *B. aristata* extracts.

Subsequently, the study focused on a non-tumor cell model in order to gather preliminary data on the possible cell function alterations induced by *B. aristata* extracts, responsible for the transformation from normal to tumor cell. Among the mechanisms involved in neoplastic transformation, the irreversible damage to DNA and a dysregulation of the cell cycle are the first to appear. In addition, the excessive generation of intracellular ROS in response to oxidative stress may lead to DNA damage and mutagenesis through the suppression and mutation of tumor suppressor genes. This phenomenon potentially promotes tumorigenesis [87,88]. The *B. aristata* extracts studied did not induce ROS production in human normal HEK293 cells, thus excluding the hypothesis of a potential effect of DNA damage mediated by excess free radicals. These data supported evidence reporting BER antioxidant activity by suppressing ROS production [89–92]. Results confirming a non-carcinogenic potential of BER also emerged from the molecular analysis in normal HEK293 cells through the evaluation of BER modulatory effects on the expression of oncogenes (*MDM2, MYC family, HRAS, MET*), or key genes in the control of cell cycle phases (*TP53, CDKN1A, CDK4*) or the apoptosis process (*BAX, BCL2*). Overall, with the exception of *c-MYC* expression being slightly upregulated by B97 20 μg/mL, a general downregulation of cell cycle genes was observed, highlighting the inhibitory effect of *B. aristata* extracts in promoting the cell cycle, and thus cell proliferation in normal cells [59,83]; moreover, no effects on the transcription of proto-oncogenes emerged, supporting data on the absence of the pro-tumoral potential of BER [16,93–95]. In addition, both the *B. aristata* extracts investigated were able to downregulate the expression of the pro-apoptotic gene *BAX*, suggesting the effects of BER in regulating at transcriptional level the apoptosis process, having no impact on apoptosis-mediated cell death in normal cells different to what was demonstrated in tumor cells [68,69,73,96]. These preliminary findings highlighted important insights into BER and PROTBERs safety in normal cells considering the crucial role of these target genes in controlling cell functions and cell homeostasis.

Summarizing, our findings allowed us to determine (i) the chemical characterization and quantification of BER and minor PROTBERs in the most marketed raw materials for food supplements; (ii) new knowledge about the pharmacokinetic and bioaccessibility properties of BER and minor PROTBERs; (iii) and finally, preliminary data on biological safety in terms of cytotoxicity and pro- or antitumoral activity.

Although the need for more research and insights into biological activity, data emerged from this study provide a broad-spectrum analysis of the most common *B. aristata* bark extracts. Overall, this evidence may represent a relevant update to better elucidate the critical points raised by the authorities related to the safety evaluation of BER-containing herbal preparations used in food supplements.

5. Conclusions

This work presents several notable strengths that deserve to be highlighted:

(i) The study addresses the bioaccessibility and safety of berberine and protoberberinoids through a highly transferrable conceptual framework. It focuses on the most widely recognized and utilized extracts available in the food supplement market, which have been chemically characterized;

(ii) Integrated In Silico and In Vitro approaches provide valuable insights. The study concurrently evaluates the bioaccessibility of the examined samples, the impact of *Berberis aristata* extracts on hepatic toxicity in the presence of CYP450 substrates, the effects of cell toxicity and cell migration on tumor cells, and the potential pro-oxidant and pro-tumoral effects in non-tumoral human cells;

(iii) The protocol employed considers different exposure times and treatment concentrations, enhancing the robustness of the In Vitro models used.

However, we aware of certain limitations of the study:

(i) In future research, we intend to expand the pharmacokinetic evaluation of the major alkaloids contained in *B. aristata* extracts, including dynamic cell absorption models;
(ii) While the current focus was on gene expression and transcriptional response to investigate early cellular response, we plan to explore the impact of berberine and protoberberinoids on human cell lines by examining upstream and downstream cell signaling pathways and protein levels related to antiproliferative activity;
(iii) Finally, we are fully aware that our findings are derived solely from In Vitro tests, and further confirmation through additional studies is necessary.

Supplementary Materials: The following supporting information can be downloaded at: https://www.mdpi.com/article/10.3390/nu16172953/s1, Figure S1: Computational analysis of jatrorrhizine (A), berberrubine (B), and palmatine (C) pharmacokinetic properties by means of SwissADME tool (http://www.swissadme.ch/); Figure S2: Genecards targets prediction for the interaction between berberine and jatrorrhizine (A), berberine and berberrubine (B), and berberine and palmatine (C) respectively; Figure S3: Analysis of target prediction by means of SwissTarget Prediction (STP) and SEA tools for berberine (A), jatrorrhizine (B), berberrubine (C), and palmatine (D).

Author Contributions: Conceptualization, M.B.; methodology, M.B. and G.R.; formal analysis and investigation, G.R., G.C., F.V. and G.B.; resources, E.M. and F.T.; writing—original draft preparation, G.R. and M.B.; review and editing, L.P., F.T. and R.B. All authors have read and agreed to the published version of the manuscript.

Funding: This research received no external funding.

Institutional Review Board Statement: Not applicable.

Informed Consent Statement: Not applicable.

Data Availability Statement: Data are available upon request to the Corresponding author upon reasonable request. Data are currently being assessed by EFSA.

Acknowledgments: This work has been carried out in the frame of the ALIFAR project of the University of Parma, funded by the Italian Ministry of University through the program "Dipartimenti di Eccellenza 2023–2027". The publication has been realized with the co-financing of the European Union—FSE-REACT-EU, PON research and Innovation 2014–2020 DM1062/2021. Authors thanks Chiara Mazzotta, University of Siena, for her precious experimental help.

Conflicts of Interest: The authors declare no conflicts of interest.

References

1. Berberine | C20H18NO4+ | CID 2353—PubChem. Available online: https://pubchem.ncbi.nlm.nih.gov/compound/2353 (accessed on 2 January 2024).
2. Neag, M.A.; Mocan, A.; Echeverría, J.; Pop, R.M.; Bocsan, C.I.; Crisan, G.; Buzoianu, A.D. Berberine: Botanical Occurrence, Traditional Uses, Extraction Methods, and Relevance in Cardiovascular, Metabolic, Hepatic, and Renal Disorders. *Front. Pharmacol.* **2018**, *9*, 343970. [CrossRef]
3. Imenshahidi, M.; Hosseinzadeh, H. Berberine and Barberry (*Berberis vulgaris*): A Clinical Review. *Phytother. Res.* **2019**, *33*, 504–523. [CrossRef]
4. European Pharmacopoeia Commission Press Release—Outcome of the 172nd Session of the European Pharmacopoeia Commission. 2022. Available online: https://www.edqm.eu/en/-/outcome-of-the-172nd-session-of-the-european-pharmacopoeia-commission-march-2022-1 (accessed on 25 June 2024).
5. Xing, L.; Zhou, X.; Li, A.H.; Li, H.J.; He, C.X.; Qin, W.; Zhao, D.; Li, P.Q.; Zhu, L.; Cao, H.L. Atheroprotective Effects and Molecular Mechanism of Berberine. *Front. Mol. Biosci.* **2021**, *8*, 762673. [CrossRef] [PubMed]
6. Yang, L.; Zhu, W.; Zhang, X.; Zhou, X.; Wu, W.; Shen, T. Efficacy and Safety of Berberine for Several Cardiovascular Diseases: A Systematic Review and Meta-Analysis of Randomized Controlled Trials. *Phytomedicine* **2023**, *112*, 154716. [CrossRef] [PubMed]
7. Cho, B.J.; Im, E.K.; Kwon, J.H.; Lee, K.H.; Shin, H.J.; Oh, J.; Kang, S.M.; Chung, J.H.; Jang, Y. Berberine Inhibits the Production of Lysophosphatidylcholine-Induced Reactive Oxygen Species and the ERK1/2 Pathway in Vascular Smooth Muscle Cells. *Mol. Cells* **2005**, *20*, 429–434. [CrossRef] [PubMed]

8. Ju, J.; Li, J.; Lin, Q.; Xu, H. Efficacy and Safety of Berberine for Dyslipidaemias: A Systematic Review and Meta-Analysis of Randomized Clinical Trials. *Phytomedicine* **2018**, *50*, 25–34. [CrossRef]
9. Li, Z.; Geng, Y.-N.; Jiang, J.-D.; Kong, W.-J. Antioxidant and Anti-Inflammatory Activities of Berberine in the Treatment of Diabetes Mellitus. *Evid. Based Complement. Alternat. Med.* **2014**, *2014*, 289264. [CrossRef]
10. Chueh, W.H.; Lin, J.Y. Berberine, an Isoquinoline Alkaloid, Inhibits Streptozotocin-Induced Apoptosis in Mouse Pancreatic Islets through down-Regulating Bax/Bcl-2 Gene Expression Ratio. *Food Chem.* **2012**, *132*, 252–260. [CrossRef] [PubMed]
11. Liang, Y.; Xu, X.; Yin, M.; Zhang, Y.; Huang, L.; Chen, R.; Ni, J. Effects of Berberine on Blood Glucose in Patients with Type 2 Diabetes Mellitus: A Systematic Literature Review and a Meta-Analysis. *Endocr. J.* **2019**, *66*, 51–63. [CrossRef] [PubMed]
12. Cao, C.; Su, M. Effects of Berberine on Glucose-Lipid Metabolism, Inflammatory Factors and Insulin Resistance in Patients with Metabolic Syndrome. *Exp. Ther. Med.* **2019**, *17*, 3009. [CrossRef] [PubMed]
13. Nazari, A.; Ghotbabadi, Z.R.; Kazemi, K.S.; Metghalchi, Y.; Tavakoli, R.; Rahimabadi, R.Z.; Ghaheri, M. The Effect of Berberine Supplementation on Glycemic Control and Inflammatory Biomarkers in Metabolic Disorders: An Umbrella Meta-Analysis of Randomized Controlled Trials. *Clin. Ther.* **2023**, *46*, e64–e72. [CrossRef] [PubMed]
14. Panigrahi, A.; Mohanty, S. Efficacy and Safety of HIMABERB® Berberine on Glycemic Control in Patients with Prediabetes: Double-Blind, Placebo-Controlled, and Randomized Pilot Trial. *BMC Endocr. Disord.* **2023**, *23*, 190. [CrossRef] [PubMed]
15. Chen, C.; Yu, Z.; Li, Y.; Fichna, J.; Storr, M. Effects of Berberine in the Gastrointestinal Tract—A Review of Actions and Therapeutic Implications. *Am. J. Chin. Med.* **2014**, *42*, 1053–1070. [CrossRef] [PubMed]
16. Zhang, L.; Wu, X.; Yang, R.; Chen, F.; Liao, Y.; Zhu, Z.; Wu, Z.; Sun, X.; Wang, L. Effects of Berberine on the Gastrointestinal Microbiota. *Front. Cell. Infect. Microbiol.* **2020**, *10*, 588517. [CrossRef] [PubMed]
17. Wang, Y.; Liu, Y.; Du, X.; Ma, H.; Yao, J. The Anti-Cancer Mechanisms of Berberine: A Review. *Cancer Manag. Res.* **2020**, *12*, 695. [CrossRef]
18. Huang, C.; Tao, L.; Wang, X.; Pang, Z. Berberine Reversed the Epithelial-Mesenchymal Transition of Normal Colonic Epithelial Cells Induced by SW480 Cells through Regulating the Important Components in the TGF-β Pathway. *J. Cell. Physiol.* **2019**, *234*, 11679–11691. [CrossRef]
19. Gong, C.; Hu, X.; Xu, Y.; Yang, J.; Zong, L.; Wang, C.; Zhu, J.; Li, Z.; Lu, D. Berberine Inhibits Proliferation and Migration of Colorectal Cancer Cells by Downregulation of GRP78. *Anticancer Drugs* **2020**, *31*, 141–149. [CrossRef]
20. Bertuccioli, A.; Cardinali, M.; Biagi, M.; Moricoli, S.; Morganti, I.; Zonzini, G.B.; Rigillo, G. Nutraceuticals and Herbal Food Supplements for Weight Loss: Is There a Prebiotic Role in the Mechanism of Action? *Microorganisms* **2021**, *9*, 2427. [CrossRef]
21. Guo, P.; Cai, C.; Wu, X.; Fan, X.; Huang, W.; Zhou, J.; Wu, Q.; Huang, Y.; Zhao, W.; Zhang, F.; et al. An Insight into the Molecular Mechanism of Berberine towards Multiple Cancer Types through Systems Pharmacology. *Front. Pharmacol.* **2019**, *10*, 857. [CrossRef]
22. Radzka, J.; Łapińska, Z.; Szwedowicz, U.; Gajewska-Naryniecka, A.; Gizak, A.; Kulbacka, J. Alternations of NF-KB Signaling by Natural Compounds in Muscle-Derived Cancers. *Int. J. Mol. Sci.* **2023**, *24*, 11900. [CrossRef]
23. Zhang, Y.; Ma, J.; Zhang, W. Berberine for Bone Regeneration: Therapeutic Potential and Molecular Mechanisms. *J. Ethnopharmacol.* **2021**, *277*, 114249. [CrossRef]
24. Noh, J.W.; Jun, M.S.; Yang, H.K.; Lee, B.C. Cellular and Molecular Mechanisms and Effects of Berberine on Obesity-Induced Inflammation. *Biomedicines* **2022**, *10*, 1739. [CrossRef] [PubMed]
25. Chen, S.; Chen, S.T.; Sun, Y.; Xu, Z.; Wang, Y.; Yao, S.Y.; Yao, W.B.; Gao, X.D. Fibroblast Growth Factor 21 Ameliorates Neurodegeneration in Rat and Cellular Models of Alzheimer's Disease. *Redox Biol.* **2019**, *22*, 101133. [CrossRef] [PubMed]
26. Wong, S.K.; Chin, K.Y.; Ima-Nirwana, S. Berberine and Musculoskeletal Disorders: The Therapeutic Potential and Underlying Molecular Mechanisms. *Phytomedicine* **2020**, *73*, 152892. [CrossRef]
27. Chen, W.; Miao, Y.Q.; Fan, D.J.; Yang, S.S.; Lin, X.; Meng, L.K.; Tang, X. Bioavailability Study of Berberine and the Enhancing Effects of TPGS on Intestinal Absorption in Rats. *AAPS PharmSciTech* **2011**, *12*, 705–711. [CrossRef]
28. He, C.Y.; Fu, J.; Shou, J.W.; Zhao, Z.X.; Ren, L.; Wang, Y.; Jiang, J.D. In Vitro Study of the Metabolic Characteristics of Eight Isoquinoline Alkaloids from Natural Plants in Rat Gut Microbiota. *Molecules* **2017**, *22*, 932. [CrossRef]
29. Zhong, F.; Chen, Y.; Chen, J.; Liao, H.; Li, Y.; Ma, Y. Jatrorrhizine: A Review of Sources, Pharmacology, Pharmacokinetics and Toxicity. *Front. Pharmacol.* **2022**, *12*, 783127. [CrossRef] [PubMed]
30. Chen, N.; Yang, X.Y.; Guo, C.E.; Bi, X.N.; Chen, J.H.; Chen, H.Y.; Li, H.P.; Lin, H.Y.; Zhang, Y.J. The Oral Bioavailability, Excretion and Cytochrome P450 Inhibition Properties of Epiberberine: An in Vivo and in Vitro Evaluation. *Drug Des. Dev. Ther.* **2018**, *12*, 57. [CrossRef] [PubMed]
31. Basera, I.A.; Girmeorcid, A.; Bhatt, V.P.; Sasteorcid, G.; Pawarorcid, S.; Hingoraniorcid, L.; Shah, M.B. Development of Validated UHPLC–PDA with ESI–MS-MS Method for Concurrent Estimation of Magnoflorine, Berbamine, Columbamine, Jatrorrhizine, Palmatine and Berberine in Berberis Aristata. *Acta Chromatogr.* **2021**, *34*, 412–421. [CrossRef]
32. Yang, N.; Sun, R.B.; Chen, X.L.; Zhen, L.; Ge, C.; Zhao, Y.Q.; He, J.; Geng, J.L.; Guo, J.H.; Yu, X.Y.; et al. In Vitro Assessment of the Glucose-Lowering Effects of Berberrubine-9-O-β-D-Glucuronide, an Active Metabolite of Berberrubine. *Acta Pharmacol. Sin.* **2017**, *38*, 351. [CrossRef]
33. Du, W.; Jin, L.; Li, L.; Wang, W.; Zeng, S.; Jiang, H.; Zhou, H. Development and Validation of a HPLC-ESI-MS/MS Method for Simultaneous Quantification of Fourteen Alkaloids in Mouse Plasma after Oral Administration of the Extract of *Corydalis yanhusuo* Tuber: Application to Pharmacokinetic Study. *Molecules* **2018**, *23*, 714. [CrossRef] [PubMed]

34. Guo, Y.; Pope, C.; Cheng, X.; Zhou, H.; Klaassen, C.D. Dose-Response of Berberine on Hepatic Cytochromes P450 MRNA Expression and Activities in Mice. *J. Ethnopharmacol.* **2011**, *138*, 111. [CrossRef]
35. Zhou, H.; Shi, R.; Ma, B.; Ma, Y.; Wang, C.; Wu, D.; Wang, X.; Cheng, N. CYP450 1A2 and Multiple UGT1A Isoforms Are Responsible for Jatrorrhizine Metabolism in Human Liver Microsomes. *Biopharm. Drug Dispos.* **2013**, *34*, 176–185. [CrossRef] [PubMed]
36. Vrba, J.; Papouskova, B.; Pyszkova, M.; Zatloukalova, M.; Lemr, K.; Ulrichova, J.; Vacek, J. Metabolism of Palmatine by Human Hepatocytes and Recombinant Cytochromes P450. *J. Pharm. Biomed. Anal.* **2015**, *102*, 193–198. [CrossRef] [PubMed]
37. Call for Data for the Scientific Opinion on the Evaluation of the Safety in Use of Plant Preparations Containing Berberine | EFSA. Available online: https://www.efsa.europa.eu/en/call/call-data-scientific-opinion-evaluation-safety-use-plant-preparations-containing-berberine (accessed on 2 January 2024).
38. Governa, P.; Manetti, F.; Miraldi, E.; Biagi, M. Effects of in Vitro Simulated Digestion on the Antioxidant Activity of Different *Camellia sinensis* (L.) Kuntze Leaves Extracts. *Eur. Food Res. Technol.* **2022**, *248*, 119–128. [CrossRef]
39. Brodkorb, A.; Egger, L.; Alminger, M.; Alvito, P.; Assunção, R.; Ballance, S.; Bohn, T.; Bourlieu-Lacanal, C.; Boutrou, R.; Carrière, F.; et al. INFOGEST Static in Vitro Simulation of Gastrointestinal Food Digestion. *Nat. Protoc.* **2019**, *14*, 991–1014. [CrossRef]
40. Daina, A.; Michielin, O.; Zoete, V. SwissADME: A Free Web Tool to Evaluate Pharmacokinetics, Drug-Likeness and Medicinal Chemistry Friendliness of Small Molecules OPEN. *Sci. Rep.* **2017**, *7*, 42717. [CrossRef]
41. Safran, M.; Rosen, N.; Twik, M.; BarShir, R.; Stein, T.I.; Dahary, D.; Fishilevich, S.; Lancet, D. The GeneCards Suite. In *Practical Guide to Life Science Databases*; Springer: Singapore, 2022; pp. 27–56. [CrossRef]
42. Yu, L.; Zhou, D.; Gao, L.; Zha, Y. Prediction of Drug Response in Multilayer Networks Based on Fusion of Multiomics Data. *Methods* **2021**, *192*, 85–92. [CrossRef]
43. Pressi, G.; Rigillo, G.; Governa, P.; Borgonetti, V.; Baini, G.; Rizzi, R.; Guarnerio, C.; Bertaiola, O.; Frigo, M.; Merlin, M.; et al. A Novel *Perilla frutescens* (L.) Britton Cell-Derived Phytocomplex Regulates Keratinocytes Inflammatory Cascade and Barrier Function and Preserves Vaginal Mucosal Integrity In Vivo. *Pharmaceutics* **2023**, *15*, 240. [CrossRef] [PubMed]
44. Pressi, G.; Bertaiola, O.; Guarnerio, C.; Barbieri, E.; Rigillo, G.; Governa, P.; Biagi, M.; Guzzo, F.; Semenzato, A. In Vitro Cell Culture of *Rhus coriaria* L.: A Standardized Phytocomplex Rich of Gallic Acid Derivatives with Antioxidant and Skin Repair Activity. *Cosmetics* **2022**, *9*, 12. [CrossRef]
45. Rigillo, G.; Ciani, M.; Benatti, C.; Blom, J.M.C.; Tascedda, F.; Pani, L.; Alboni, S.; Brunello, N. Vortioxetine Attenuates Neuroinflammation by Modulating the NOD-like Receptor Family Pyrin Domain Containing 3 Inflammasome Activation in Microglia: Implications for Cognitive Function. *Neurosci. Appl.* **2023**, *2*, 103728. [CrossRef]
46. Rigillo, G.; Belluti, S.; Campani, V.; Ragazzini, G.; Ronzio, M.; Miserocchi, G.; Bighi, B.; Cuoghi, L.; Mularoni, V.; Zappavigna, V.; et al. The NF-Y Splicing Signature Controls Hybrid EMT and ECM-Related Pathways to Promote Aggressiveness of Colon Cancer. *Cancer Lett.* **2023**, *567*, 216262. [CrossRef]
47. Rigillo, G.; Basile, V.; Belluti, S.; Ronzio, M.; Sauta, E.; Ciarrocchi, A.; Latella, L.; Saclier, M.; Molinari, S.; Vallarola, A.; et al. The Transcription Factor NF-Y Participates to Stem Cell Fate Decision and Regeneration in Adult Skeletal Muscle. *Nat. Commun.* **2021**, *12*, 6013. [CrossRef] [PubMed]
48. Ciani, M.; Rigillo, G.; Benatti, C.; Pani, L.; Blom, J.M.C.; Brunello, N.; Tascedda, F.; Alboni, S. Time- and Region-Specific Effect of Vortioxetine on Central LPS-Induced Regulation of NLRP3 Inflammasome. *Curr. Neuropharmacol.* **2024**, *22*, 1–13. [CrossRef]
49. Ichim, M.C.; Booker, A. Chemical Authentication of Botanical Ingredients: A Review of Commercial Herbal Products. *Front. Pharmacol.* **2021**, *12*, 666850. [CrossRef]
50. Danoun, S.; Balayssac, S.; Gilard, V.; Martino, R.; Malet-Martino, M. Quality Evaluation of Berberine Food Supplements with High-Field and Compact 1H NMR Spectrometers. *J. Pharm. Biomed. Anal.* **2023**, *223*, 115161. [CrossRef] [PubMed]
51. Maharjan, B.; Payne, D.T.; Ferrarese, I.; Giovanna Lupo, M.; Kumar Shrestha, L.; Hill, J.P.; Ariga, K.; Rossi, I.; Sharan Shrestha, S.; Panighel, G.; et al. Evaluation of the Effects of Natural Isoquinoline Alkaloids on Low Density Lipoprotein Receptor (LDLR) and Proprotein Convertase Subtilisin/Kexin Type 9 (PCSK9) in Hepatocytes, as New Potential Hypocholesterolemic Agents. *Bioorg. Chem.* **2022**, *121*, 105686. [CrossRef]
52. Lee, S.; Lim, H.J.; Park, J.H.; Lee, K.S.; Jang, Y.; Park, H.Y. Berberine-Induced LDLR up-Regulation Involves JNK Pathway. *Biochem. Biophys. Res. Commun.* **2007**, *362*, 853–857. [CrossRef] [PubMed]
53. Katiyar, S.K.; Meeran, S.M.; Katiyar, N.; Akhtar, S. P53 Cooperates Berberine-Induced Growth Inhibition and Apoptosis of Non-Small Cell Human Lung Cancer Cells in Vitro and Tumor Xenograft Growth in Vivo. *Mol. Carcinog.* **2009**, *48*, 24–37. [CrossRef] [PubMed]
54. Hsu, W.H.; Hsieh, Y.S.; Kuo, H.C.; Teng, C.Y.; Huang, H.I.; Wang, C.J.; Yang, S.F.; Liou, Y.S.; Kuo, W.H. Berberine Induces Apoptosis in SW620 Human Colonic Carcinoma Cells through Generation of Reactive Oxygen Species and Activation of JNK/P38 MAPK and FasL. *Arch. Toxicol.* **2007**, *81*, 719–728. [CrossRef] [PubMed]
55. Cheng, Z.; Kang, C.; Che, S.; Su, J.; Sun, Q.; Ge, T.; Guo, Y.; Lv, J.; Sun, Z.; Yang, W.; et al. Berberine: A Promising Treatment for Neurodegenerative Diseases. *Front. Pharmacol.* **2022**, *13*, 845591. [CrossRef] [PubMed]
56. Chen, F.L.; Yang, Z.H.; Liu, Y.; Li, L.X.; Liang, W.C.; Wang, X.C.; Zhou, W.B.; Yang, Y.H.; Hu, R.M. Berberine Inhibits the Expression of TNFalpha, MCP-1, and IL-6 in AcLDL-Stimulated Macrophages through PPARgamma Pathway. *Endocrine* **2008**, *33*, 331–337. [CrossRef] [PubMed]

57. Li, J.; Cao, B.; Liu, X.; Fu, X.; Xiong, Z.; Chen, L.; Sartor, O.; Dong, Y.; Zhang, H. Berberine Suppresses Androgen Receptor Signaling in Prostate Cancer. *Mol. Cancer Ther.* **2011**, *10*, 1346. [CrossRef] [PubMed]
58. Jantová, S.; čipák, L.; čerňáková, M.; Košt'álová, D. Effect of Berberine on Proliferation, Cell Cycle and Apoptosis in HeLa and L1210 Cells. *J. Pharm. Pharmacol.* **2010**, *55*, 1143–1149. [CrossRef] [PubMed]
59. Li, Z.; Chen, M.; Wang, Z.; Fan, Q.; Lin, Z.; Tao, X.; Wu, J.; Liu, Z.; Lin, R.; Zhao, C. Berberine Inhibits RA-FLS Cell Proliferation and Adhesion by Regulating RAS/MAPK/FOXO/HIF-1 Signal Pathway in the Treatment of Rheumatoid Arthritis. *Bone Jt. Res.* **2023**, *12*, 91–102. [CrossRef]
60. Li, J.; Liu, F.; Jiang, S.; Liu, J.; Chen, X.; Zhang, S.; Zhao, H. Berberine Hydrochloride Inhibits Cell Proliferation and Promotes Apoptosis of Non-Small Cell Lung Cancer via the Suppression of the MMP2 and Bcl-2/Bax Signaling Pathways. *Oncol. Lett.* **2018**, *15*, 7409. [CrossRef]
61. Lo, S.N.; Wang, C.W.; Chen, Y.S.; Huang, C.C.; Wu, T.S.; Li, L.A.; Lee, I.J.; Ueng, Y.F. Berberine Activates Aryl Hydrocarbon Receptor but Suppresses CYP1A1 Induction through MiR-21-3p Stimulation in MCF-7 Breast Cancer Cells. *Molecules* **2017**, *22*, 1847. [CrossRef]
62. Wei, W.; Yao, J.X.; Zhang, T.T.; Wen, J.Y.; Zhang, Z.; Luo, Y.M.; Cao, Y.; Li, H. Network Pharmacology Reveals That Berberine May Function against Alzheimer's Disease via the AKT Signaling Pathway. *Front. Neurosci.* **2023**, *17*, 1059496. [CrossRef]
63. Zhao, L.; Liu, S.; Wang, M.; Zhi, M.; Geng, X.; Hou, C.; Wang, W.; Zhao, D. Berberine Restored Nitrergic and Adrenergic Function in Mesenteric and Iliac Arteries from Streptozotocin-Induced Diabetic Rats. *J. Ethnopharmacol.* **2019**, *244*, 112140. [CrossRef] [PubMed]
64. Mak, S.; Luk, W.W.K.; Cui, W.; Hu, S.; Tsim, K.W.K.; Han, Y. Synergistic Inhibition on Acetylcholinesterase by the Combination of Berberine and Palmatine Originally Isolated from Chinese Medicinal Herbs. *J. Mol. Neurosci.* **2014**, *53*, 511–516. [CrossRef]
65. Campbell, K.J.; Tait, S.W.G. Targeting BCL-2 Regulated Apoptosis in Cancer. *Open Biol.* **2018**, *8*, 180002. [CrossRef] [PubMed]
66. Qian, S.; Wei, Z.; Yang, W.; Huang, J.; Yang, Y.; Wang, J. The Role of BCL-2 Family Proteins in Regulating Apoptosis and Cancer Therapy. *Front. Oncol.* **2022**, *12*, 985363. [CrossRef]
67. Singh, R.; Letai, A.; Sarosiek, K. Regulation of Apoptosis in Health and Disease: The Balancing Act of BCL-2 Family Proteins. *Nat. Rev. Mol. Cell Biol.* **2019**, *20*, 175–193. [CrossRef]
68. Li, J.; Li, O.; Kan, M.; Zhang, M.; Shao, D.; Pan, Y.; Zheng, H.; Zhang, X.; Chen, L.; Liu, S. Berberine Induces Apoptosis by Suppressing the Arachidonic Acid Metabolic Pathway in Hepatocellular Carcinoma. *Mol. Med. Rep.* **2015**, *12*, 4572–4577. [CrossRef]
69. Palma, T.V.; Lenz, L.S.; Bottari, N.B.; Pereira, A.; Schetinger, M.R.C.; Morsch, V.M.; Ulrich, H.; Pillat, M.M.; de Andrade, C.M. Berberine Induces Apoptosis in Glioblastoma Multiforme U87MG Cells via Oxidative Stress and Independent of AMPK Activity. *Mol. Biol. Rep.* **2020**, *47*, 4393–4400. [CrossRef]
70. Kuo, C.L.; Chou, C.C.; Yung, B.Y.M. Berberine Complexes with DNA in the Berberine-Induced Apoptosis in Human Leukemic HL-60 Cells. *Cancer Lett.* **1995**, *93*, 193–200. [CrossRef] [PubMed]
71. Liu, Z.; Liu, Q.; Xu, B.; Wu, J.; Guo, C.; Zhu, F.; Yang, Q.; Gao, G.; Gong, Y.; Shao, C. Berberine Induces P53-Dependent Cell Cycle Arrest and Apoptosis of Human Osteosarcoma Cells by Inflicting DNA Damage. *Mutat. Res.-Fundam. Mol. Mech. Mutagen.* **2009**, *662*, 75–83. [CrossRef] [PubMed]
72. Xie, J.; Xu, Y.; Huang, X.; Chen, Y.; Fu, J.; Xi, M.; Wang, L. Berberine-Induced Apoptosis in Human Breast Cancer Cells Is Mediated by Reactive Oxygen Species Generation and Mitochondrial-Related Apoptotic Pathway. *Tumor Biol.* **2015**, *36*, 1279–1288. [CrossRef]
73. Liu, L.; Fan, J.; Ai, G.; Liu, J.; Luo, N.; Li, C.; Cheng, Z. Berberine in Combination with Cisplatin Induces Necroptosis and Apoptosis in Ovarian Cancer Cells. *Biol. Res.* **2019**, *52*, 37. [CrossRef]
74. Yao, Z.; Wan, Y.; Li, B.; Zhai, C.; Yao, F.; Kang, Y.; Liu, Q.; Lin, D. Berberine Induces Mitochondrial-Mediated Apoptosis and Protective Autophagy in Human Malignant Pleural Mesothelioma NCI-H2452 Cells. *Oncol. Rep.* **2018**, *40*, 3603–3610. [CrossRef]
75. Li, W.; Li, D.; Kuang, H.; Feng, X.; Ai, W.; Wang, Y.; Shi, S.; Chen, J.; Fan, R. Berberine Increases Glucose Uptake and Intracellular ROS Levels by Promoting Sirtuin 3 Ubiquitination. *Biomed. Pharmacother.* **2020**, *121*, 109563. [CrossRef] [PubMed]
76. Feng, P.-F.; Zhu, L.-X.; Jie, J.; Yang, P.-X.; Chen, X. The Intracellular Mechanism of Berberine-Induced Inhibition of CYP3A4 Activity. *Curr. Pharm. Des.* **2021**, *27*, 4179–4185. [CrossRef]
77. Tian, E.; Sharma, G.; Dai, C. Neuroprotective Properties of Berberine: Molecular Mechanisms and Clinical Implications. *Antioxidants* **2023**, *12*, 1883. [CrossRef]
78. Huang, Q.; Ji, D.; Tian, X.; Ma, L.; Sun, X. Berberine Inhibits Erastin-Induced Ferroptosis of Mouse Hippocampal Neuronal Cells Possibly by Activating the Nrf2-HO-1/GPX4 Pathway. *J. South. Med. Univ.* **2022**, *42*, 937–943. [CrossRef]
79. Gu, Y.; Zhou, Z. Berberine Inhibits the Proliferation, Invasion and Migration of Endometrial Stromal Cells by Downregulating MiR-429. *Mol. Med. Rep.* **2021**, *23*, 588517. [CrossRef] [PubMed]
80. Li, Q.; Qing, L.; Xu, W.; Yang, Y.; You, C.; Lao, Y.; Dong, Z. Berberine Affects the Proliferation, Migration, Invasion, Cell Cycle, and Apoptosis of Bladder Cancer Cells T24 and 5637 by Down-Regulating the HER2/PI3K/AKT Signaling Pathway. *Arch. Esp. Urol.* **2023**, *76*, 152–160. [CrossRef]
81. Burkina, V.; Rasmussen, M.K.; Pilipenko, N.; Zamaratskaia, G. Comparison of Xenobiotic-Metabolising Human, Porcine, Rodent, and Piscine Cytochrome P450. *Toxicology* **2017**, *375*, 10–27. [CrossRef]

82. Khan, M.; Giessrigl, B.; Vonach, C.; Madlener, S.; Prinz, S.; Herbaceck, I.; Hölzl, C.; Bauer, S.; Viola, K.; Mikulits, W.; et al. Berberine and a Berberis Lycium Extract Inactivate Cdc25A and Induce α-Tubulin Acetylation That Correlate with HL-60 Cell Cycle Inhibition and Apoptosis. *Mutat. Res./Fundam. Mol. Mech. Mutagen.* **2010**, *683*, 123–130. [CrossRef] [PubMed]
83. Liu, J.F.; Lai, K.C.; Peng, S.F.; Maraming, P.; Huang, Y.P.; Huang, A.C.; Chueh, F.S.; Huang, W.W.; Chung, J.G. Berberine Inhibits Human Melanoma A375.S2 Cell Migration and Invasion via Affecting the FAK, UPA, and NF-KB Signaling Pathways and Inhibits PLX4032 Resistant A375.S2 Cell Migration in Vitro. *Molecules* **2018**, *23*, 2019. [CrossRef]
84. Yang, L.J.; He, J.B.; Jiang, Y.; Li, J.; Zhou, Z.W.; Zhang, C.; Tao, X.; Chen, A.F.; Peng, C.; Xie, H.H. Berberine Hydrochloride Inhibits Migration Ability via Increasing Inducible NO Synthase and Peroxynitrite in HTR-8/SVneo Cells. *J. Ethnopharmacol.* **2023**, *305*, 116087. [CrossRef]
85. Tarawneh, N.; Hamadneh, L.; Abu-Irmaileh, B.; Shraideh, Z.; Bustanji, Y.; Abdalla, S. Berberine Inhibited Growth and Migration of Human Colon Cancer Cell Lines by Increasing Phosphatase and Tensin and Inhibiting Aquaporins 1, 3 and 5 Expressions. *Molecules* **2023**, *28*, 3823. [CrossRef] [PubMed]
86. Srinivas, U.S.; Tan, B.W.Q.; Vellayappan, B.A.; Jeyasekharan, A.D. ROS and the DNA Damage Response in Cancer. *Redox Biol.* **2019**, *25*, 101084. [CrossRef] [PubMed]
87. Sarmiento-Salinas, F.L.; Perez-Gonzalez, A.; Acosta-Casique, A.; Ix-Ballote, A.; Diaz, A.; Treviño, S.; Rosas-Murrieta, N.H.; Millán-Perez-Peña, L.; Maycotte, P. Reactive Oxygen Species: Role in Carcinogenesis, Cancer Cell Signaling and Tumor Progression. *Life Sci.* **2021**, *284*, 119942. [CrossRef] [PubMed]
88. Kassab, R.B.; Vasicek, O.; Ciz, M.; Lojek, A.; Perecko, T. The Effects of Berberine on Reactive Oxygen Species Production in Human Neutrophils and in Cell-Free Assays. *Interdiscip. Toxicol.* **2017**, *10*, 61. [CrossRef]
89. Lv, X.X.; Yu, X.H.; Wang, H.D.; Yan, Y.X.; Wang, Y.P.; Lu, D.X.; Qi, R.B.; Hu, C.F.; Li, H.M. Berberine Inhibits Norepinephrine-Induced Apoptosis in Neonatal Rat Cardiomyocytes via Inhibiting ROS-TNF-α-Caspase Signaling Pathway. *Chin. J. Integr. Med.* **2013**, *19*, 424–431. [CrossRef]
90. Sun, Y.; Yuan, X.; Zhang, F.; Han, Y.; Chang, X.; Xu, X.; Li, Y.; Gao, X. Berberine Ameliorates Fatty Acid-Induced Oxidative Stress in Human Hepatoma Cells. *Sci. Rep.* **2017**, *7*, 11340. [CrossRef] [PubMed]
91. Dong, S.F.; Yasui, N.; Negishi, H.; Kishimoto, A.; Sun, J.N.; Ikeda, K. Increased Oxidative Stress in Cultured 3T3-L1 Cells Was Attenuated by Berberine Treatment. *Nat. Prod. Commun.* **2015**, *10*, 895–897. [CrossRef] [PubMed]
92. Wen, L.; Han, Z.; Li, J.; Du, Y. C-MYC and HIF1α Promoter G-Quadruplexes Dependent Metabolic Regulation Mechanism of Berberine in Colon Cancer. *J. Gastrointest. Oncol.* **2022**, *13*, 1152–1168. [CrossRef] [PubMed]
93. Rauf, A.; Abu-Izneid, T.; Khalil, A.A.; Imran, M.; Shah, Z.A.; Bin Emran, T.; Mitra, S.; Khan, Z.; Alhumaydhi, F.A.; Aljohani, A.S.M.; et al. Berberine as a Potential Anticancer Agent: A Comprehensive Review. *Molecules* **2021**, *26*, 7368. [CrossRef]
94. Almatroodi, S.A.; Alsahli, M.A.; Rahmani, A.H. Berberine: An Important Emphasis on Its Anticancer Effects through Modulation of Various Cell Signaling Pathways. *Molecules* **2022**, *27*, 5889. [CrossRef] [PubMed]
95. Ho, Y.; Lu, C.-C.; Yang, J.-S.; Chiang, J.-H.; Li, T.-C.; Ip, S.; Hsia, T.; Liao, C.; Lin, J.-G.; Wood, W.; et al. Berberine Induced Apoptosis via Promoting the Expression of Caspase-8, -9 and -3, Apoptosis-Inducing Factor and Endonuclease G in SCC-4 Human Tongue Squamous Carcinoma Cancer Cells. *Anticancer Res.* **2009**, *29*, 4063–4070. [PubMed]
96. Kim, D.-Y.; Kim, S.-H.; Cheong, H.-T.; Ra, C.-S.; Rhee, K.-J.; Jung, B.D. Berberine Induces P53-Dependent Apoptosis through Inhibition of DNA Methyltransferase3b in Hep3B Cells. *Korean J. Clin. Lab. Sci.* **2020**, *52*, 69–77. [CrossRef]

nutrients

MDPI

Article

Effect of Low-Dose Mulberry Fruit Extract on Postprandial Glucose and Insulin Responses: A Randomized Pilot Trial in Individuals with Type 2 Diabetes

David J. Mela [1,*,†], Marjan Alssema [1,‡], Harry Hiemstra [1,†], Anne-Roos Hoogenraad [1] and Tanvi Kadam [2]

1 Unilever Foods Innovation Centre Wageningen, 6708 WH Wageningen, The Netherlands
2 Unilever Industries Private Limited, Mumbai 400099, India
* Correspondence: djmela@djmela.eu
† Retired.
‡ Current affiliation: Health Council of The Netherlands, 2594 AV Den Haag, The Netherlands.

Abstract: Adding mulberry fruit extract (MFE) to carbohydrate-rich meals can reduce postprandial glucose (PPG) and insulin (PPI) responses in healthy individuals. This pilot study assessed the acute postprandial effects of low doses of MFE in individuals with type 2 diabetes. In a randomized cross-over (within-subjects) design, 24 unmedicated adult males and females with type 2 diabetes (mean [SD] age 51.0 [9.3] yr, BMI 27.5 [3.9] kg/m^2) consumed meals with 0 (control), 0.37, and 0.75 g of MFE added to ~50 g of available carbohydrates from rice. Primary and secondary outcomes were the PPG 2 hr positive incremental area under the curve and the corresponding PPI. Results were reported as mean differences from the control meal with 95% CI. Relative to control, 0.37 and 0.75 g of MFE reduced the mean 2 hr PPG by 8.2% (−20.8 to 6.6%) and 22.4% (−38.6 to −1.9%), respectively, and reduced PPI by 9.6% (−20.7 to 3.0%) and 17.5% (−27.9 to −5.7%). There were no indications of adverse events or gastrointestinal discomfort. MFE additions also led to dose-related reductions in glucose peak and glucose swing. At these levels, MFE appears to dose-dependently reduce acute PPG and PPI in individuals with type 2 diabetes and may be a feasible dietary approach to help attenuate glycemic exposures.

Keywords: alpha-glucosidase; 1-deoxynojirimycin; glycemic control; starch; rice

1. Introduction

The key treatment target for the management of diabetes is the improvement of glycemic control, which includes not only fasting glucose levels but also postprandial glucose (PPG) responses [1,2]. Drugs that target PPG by slowing the digestion of carbohydrates have been shown to be beneficial for glycemic control, as well as for reducing the risk of diabetes onset in individuals with pre-diabetes [3–5]. Dietary guidelines for diabetes include recommendations to help reduce and manage PPG, mainly by controlling the quantity and quality of carbohydrate-containing foods. A possible additional dietary approach is the use of food components that specifically slow the digestion of glycemic carbohydrates in foods and hence the rate of appearance of glucose in blood [6].

Mulberry products are a source of iminosugar 1-deoxynojirimycin (DNJ), which acts as an alpha-glucosidase inhibitor [7]. The inhibition of alpha-glucosidase slows the final step in the digestion of dietary carbohydrates and can thus reduce the rate of uptake and appearance of glucose in blood. A range of naturally-occurring and synthetic iminosugar-based molecules show alpha-glucosidase inhibition, making this the basis for a class of existing drugs (e.g., acarbose and miglitol) and, potentially, also new drugs for managing diabetes [8,9]. We have shown in a series of controlled trials that a well-characterized mulberry fruit extract (MFE) containing 0.5% DNJ reduces both PPG and post-prandial insulin (PPI) responses to test meals in healthy human subjects without diabetes [10–13].

In those trials, MFE was effective and well-tolerated in tested doses ranging from 0.37 to 1.5 g of MFE (~1.85 to 7.5 mg of DNJ) added to meals containing ~50 g of available carbohydrates from rice or wheat. Importantly, we also confirmed that MFE in this dose range reduces the rate but not the amount of glucose uptake, with no indications of carbohydrate malabsorption or adverse gastrointestinal symptoms [10,11].

In contrast, most research testing mulberry products for glycemic control have used mulberry leaf extract (MLE) with 2–10 times higher extract doses and DNJ contents [14,15]. That includes a small number of studies indicating the potential benefits of mulberry products, mainly MLE, in individuals with pre-diabetes or impaired fasting glucose or glucose tolerance [16]. A number of studies on acute PPG responses in those populations have reported some indication of efficacy for PPG lowering using MLE doses from 0.4 to over 3 g, containing (where reported) 6 to 25 mg of DNJ [17–21], although one study found no significant effect at these levels [22]. Most of these trials did not identify adverse effects, although significantly increased breath hydrogen was reported by Mudra et al. (1 g of MLE, DNJ not specified) and Nakamura et al. (3.3 g of MLE, 25 mg of DNJ) [19,20].

The present study was intended as a pilot test to assess the potential effects of low doses of a well-characterized MFE on acute PPG and PPI responses in individuals diagnosed with type II diabetes and not on glucose-lowering medications. MFE at the two lowest effective dose levels (0.37 and 0.75 g of MFE containing ~2–4 mg of DNJ) in our previous research [10] were added to ~50 g of available carbohydrates from white rice and compared to the rice alone with no MFE (control).

2. Materials and Methods

2.1. General

This was a randomized, double-blind, multi-center, 3-period-balanced order cross-over (within-subject) trial testing two doses of MFE added to boiled rice, compared to rice alone (control), in individuals with type 2 diabetes. Rice was used as a test food because it is a major carbohydrate staple in this population, and it allowed for results to be directly compared to our prior research using MFE added to rice in subjects without diabetes [10]. The pre-registered primary outcome was the effect of the MFE additions on venous PPG, expressed as the percent difference in the positive incremental area under the curve over 2 hr (+$iAUC_{2hr}$), relative to control. Secondary outcomes were the corresponding effects on the PPI total area under the curve ($tAUC_{2hr}$) and the measures of safety and tolerance. Exploratory outcomes were the 3 and 4 hr PPG (+$iAUC_{3hr}$ and +$iAUC_{4hr}$) and PPI ($tAUC_{3hr}$, $tAUC_{4hr}$) responses, the peak glucose level (C_{max}), the glucose swing (amplitude of response, defined as the maximum–minimum postprandial concentration [C_{max}–C_{min}]), and changes in mean urine glucose levels over 4 hr following the consumption of the rice meals.

The study was not planned for formal statistical hypothesis testing (testing for statistical significance), although inferences may be made from the reported means and 95% confidence intervals corrected for multiple comparisons. A formal power (sample size) calculation was not possible because essential information, such as variability in glucose AUC in the intended study population, was not available. For typical pilot/exploratory studies, a minimal number of 12 subjects is recommended as a 'rule of thumb', based on feasibility and the diminishing gains in precision with greater sample sizes [23]. A 15% reduction relative to the control was considered a priori to be a desired and physiologically meaningful target effect size. That benchmark roughly corresponded to the 15 unit difference between cut-offs for 'low' (55) and 'high' (70) glycemic index values, which is associated with health benefits in healthy populations, as well as in individuals with diabetes [24,25]. The study was planned for a sample size of 24 subjects, as we anticipated a larger variability of the glucose and insulin profiles within this population with diabetes, relative to our previous research in individuals without diabetes.

The trial was prospectively registered at clinicaltrials.gov with the identifier NCT02256332. The clinical phase was executed between 16 February and 16 July 2015 at three study sites in India: Ashirwad Hospital and Research Centre, Ulhasnagar, Thane (site 1); Bangalore Clinisearch, Bangalore (site 2); and Therapeutic Drug Monitoring Laboratory, Mumbai (site 3). Lambda Therapeutics Research Ltd. (LTRL), Ahmadabad, India acted as the central laboratory and data management center. The sponsoring company provided study materials but had no role in participant contact, study execution, or outcome measurement and recording.

2.2. Ethical Approval

The trial was conducted in compliance with the Declaration of Helsinki, and the protocol and informed consent forms were approved as Protocol FDS-NAA-1633 by the Ashirwad Ethics Committee (site 1) on 6 November 2014, the Medisys Clinisearch Ethical Review Board (site 2) on 17 January 2015, and the Therapeutic Drug Monitoring Laboratory Institutional Ethics Committee (site 3) on 23 March 2015.

2.3. Participants and Allocation to Treatments

Participants were otherwise healthy adults with type 2 diabetes not treated with drugs. The full in- and exclusion criteria are given in Supplementary Material, Table S1. In brief, eligible individuals were males and females age 20–65, with a BMI of 18–35, with type 2 diabetes (HbA1c $\geq$ 6.5%) controlled only through diet and exercise, and who were not treated with glucose-lowering drugs in the preceding three months.

Individuals were provisionally invited to participate by physician investigators at each study site on the basis of medical records and the study inclusion/exclusion criteria. At an initial screening visit, potential participants were given verbal and written explanations and the opportunity to inquire about the details of the research. They were informed of their right to withdraw from the study at any time and gave written informed consent before the start of any protocol-specific procedures. All explanations and procedures were conducted in the native language of participants. Individuals consenting to participate in the study underwent further screening procedures at the study site. This included verification of basic personal and medical history information and a physical examination, including anthropometric measures, urine drug screening, and a blood sample collection for the analysis of HbA1c, hemoglobin, lipids, and routine blood chemistry. Eligibility of the subjects to enter the study was then determined on the basis of the results of screening.

Subjects all received one of each of the test products at test sessions over 3 weeks: control (rice alone, no MFE); rice + 0.37 g of MFE; and rice + 0.75 g of MFE in a balanced-order design. The dose levels were chosen based on efficacy in previous dose-response trials with this same specific extract. With 3 treatment arms (test products), there were 6 possible treatment order sequences. In order to achieve the balanced-order design, equal numbers of the 24 eligible individuals were assigned to one of these six possible treatment sequences using computer-generated random allocations (Supplementary Material, Figure S1). Anyone who dropped out before the first study product administration would be replaced. Anyone dropping out after participating in any of the study product administration periods would not be replaced. The randomization schedule was kept under controlled access by an individual not involved in the study and was unavailable to any personnel who could have an impact on the outcome of the study, e.g., recording of clinical laboratory or other subject data or the collection or evaluation of adverse events. The treatment codes were only broken after the completion of a blind review and a hard lock of the database.

2.4. Source and Characterization of MFE

The MFE (batch No. MF-DC-KQ-111207 Draco Natural Products Inc., San Jose, CA, USA) contained, by weight, 0.5% DNJ (~1.85 and 3.75 mg of DNJ, respectively, in the 0.37 and 0.75 g of MFE dose levels). This is a commercially available aqueous extract produced using a proprietary process and standardized for the DNJ content. The DNJ content and in vitro bioactivity (alpha-glucosidase inhibition), as well as in vivo efficacy of this specific batch of MFE, have previously been confirmed and reported [10,13]. The

MFE was packaged in pre-weighed, individually coded aluminum sachets (Pharamivize, Mariakerke, Belgium) containing either 0 (control), 0.37, or 0.75 g of MFE plus mannitol to bring the total weight of the sachet contents to 1 g.

2.5. Test Meals

Each test meal consisted of a serving of boiled rice with 0, 0.37, or 0.75 g of MFE added. A single serving of rice was prepared using 64 g of raw Sona Maroori (Sona Masuri) rice containing ~50 g of available carbohydrates, sourced and prepared as previously described [12]. Each portion of rice was prepared in a rice cooker with the addition of 140 mL of water. Sachets containing the control or MFE were then stirred into each serving of prepared rice. Participants and staff serving the rice were blind to the presence or dose of MFE, which, at these levels, has negligible effects on sensory attributes.

2.6. Test Day Procedures and Data Collection

Subjects participated in three test days separated by 5–7 days. They arrived at the test facility at around 18.00 on the evenings prior to test days and stayed at the facility at least 4 hr after the administration of the test meals. Subjects were asked to avoid strenuous physical activity and the consumption of alcohol for at least 24 hr and were fasted for at least 10 hr prior to the start of the test meals. They were given identical, standardized evening meals of a fixed quantity and not permitted to consume any other food or beverages except water, which was allowed up to 1 hr prior to test meals.

On test days, subjects consumed the test meals immediately after preparation, together with 350 mL of water, as a morning meal. The moment the first mouthful of rice was swallowed was recorded as t = 0 (0 hr, 0 min). Subjects were instructed to consume the study product within 15 min. If a subject was not able to finish the preparation within this time frame, the 15 min blood sample was taken, and they continued consuming the study product immediately after blood sampling. Subjects did not consume any other food and were allowed, at a maximum, an additional 500 mL of water until the last blood sample was taken and a gastrointestinal discomfort questionnaire was completed at t = 4 hr. The actual quantity of water consumed by each subject was recorded on their first test day, and it was the same amount permitted on subsequent test days.

For the analysis of plasma glucose and serum insulin, two consecutive baseline blood samples were collected with a maximum gap of five minutes and within a period of 15 min before the start of the test meal. Following the start of the test meal at t = 0 min, blood samples were collected at t = 15, 30, 45, 60, 90, 120, 150, 180, 210, and 240 min for analyses.

Venous blood samples (6 mL) were collected from a valve on an indwelling cannula placed in the forearm (antecubital vein), which was kept patent by injections of 0.5 mL of normal saline solution. At each timepoint, the first 0.5 mL of blood was collected into a 2 mL syringe and discarded to avoid saline contamination. Blood was subsequently collected into vacuumized sample collection tubes attached with a Luer adapter to the cannula. If the cannula was blocked or there was difficulty in drawing blood through the cannula, blood samples could be taken either directly from the cannula into a 10 mL syringe or by a fresh vein puncture using a 22-gauge needle.

Subjects completed a gastrointestinal discomfort questionnaire 30 min prior to and at 240 min after the consumption of the test meals. The questionnaire asked whether subjects had experienced any flatulence, nausea, bloating, or bowel pain, with each symptom separately rated as none, mild, moderate, or severe.

For the analysis of urine glucose, a baseline urine sample was collected approximately 30–45 min before each test meal. Urine subsequently produced by each subject was collected in pots during the 4 hr following the consumption of test products and immediately refrigerated. The last urine sample was collected at approximately 4 hr after the start of the test meal when other procedures were completed.

2.7. Blood and Urine Sample Handling and Analyses

Duplicate aliquots of plasma (for glucose) and serum (for insulin) were obtained from 2 and 4 mL of blood, respectively. Samples were centrifuged at 18–25 °C within 60 min of collection for serum and within 45 min for plasma at 2500–3000 rpm for 10 min. Proper clot formation was ensured before centrifugation for serum separation. The samples were frozen and stored at -22 ± 5 °C within 15 min of centrifugation. The volumes of individual baseline urine samples and of the pooled urine samples following the consumption of the test products were measured, and two 6 mL aliquots of each were frozen for later analyses. One set of all samples was shipped, frozen, to LTRL for analyses, and a duplicate set was retained at each study site in case of loss or the need for re-analysis.

Glucose in plasma and urine was analyzed using a Vitros 5,1 FS Chemistry System analyzer (Ortho Clinical Diagnostics, Inc.; Raritan, NJ, USA). Insulin was measured by electrochemiluminescence on a COBAS e411 immunoassay analyzer (Roche; Basel, Switzerland).

2.8. Adverse Event Recording

Adverse events were defined as mild, moderate, or severe, according to the need for treatment and the level of interference with normal daily activities. The possible relationship of any adverse event to the study meals or procedure was defined as unrelated, unlikely, possibly, probably, or definitely related, according to the criteria, including the association with time and the likelihood of alternative explanations.

2.9. Statistical Analyses

Statistical analyses were carried out according to a pre-specified plan, and no interim analyses were planned or performed. The primary endpoint was calculated with a linear mixed model using Log(+$iAUC_{0-120min}$) as the response. The models always included baseline, subject_baseline, treatment, and treatment sequence order as predictors. Baseline was the mean baseline value for that visit for that subject; subject_baseline was the mean baseline score over all visits, and it was included to avoid possible bias in the estimates of the treatment effect due to the use of a mixed model and the inclusion of a different baseline value at each visit for the subject. The error term of the model was assumed to be normally distributed. Other predictors, such as body weight, gender, and visit (a categorical variable identifying the number of test day visits), could be included in the model based on statistical relevance. The model-derived estimated differences in treatment effects (obtained on a log scale) were back-transformed into a percent change and its associated confidence interval.

An analogous statistical model was used for the secondary outcome PPI $tAUC_{2h}$ and the corresponding exploratory results for PPG and PPI over 3 and 4 hr. For these PPG and PPI outcomes and the exploratory endpoint C_{max}, the percent change at each dose of MFE was calculated relative to the control rice, with no MFE as a reference. The other exploratory endpoints, glucose swing (C_{max}–C_{min}) and pre- vs. post-meal glucose concentration in urine, were determined for each treatment and reported as absolute differences for each MFE treatment relative to the control.

This was a pilot study that was not planned or powered for formal statistical hypothesis testing (determination of *p*-values). The relevant results are therefore presented as the size of the estimated treatment effect: mean and 95% confidence intervals (CI) were used as measures of its reliability using a Dunnett correction for multiple comparisons with the control. Results where the 95% CI for the difference between an MFE treatment and control did not include a null effect (zero) would therefore be analogous to $p < 0.05$.

3. Results

3.1. Analysis Population

Subject baseline characteristics are shown in Table 1. Of the 24 individuals entering the trial, 22 completed all treatments. One subject provided no usable treatment-related data, and one dropped out after their first test session (see Consolidated Standards of Reporting

Trials [CONSORT] subject flow diagram, Supplementary Material Figure S1). During blind review, high baseline values for both plasma and urinary glucose (exceeding 10 mmol/L and 250 ug/mL, respectively) were observed from one subject at all visits and from a further subject at one visit. These values were attributed to subjects not being fasted, and it was judged that the glucose and insulin data from those subjects/visits should be excluded from the statistical analyses. Available data from all other participants and visits were used, and no distinction was made between the intention-to-treat and per-protocol analyses.

Table 1. Subject baseline characteristics (N = 24; 11 male, 13 female).

	Mean	SD	Range
Age, yr	51	9.3	34–63
Weight, kg	69.7	20.3	47.6–88.2
BMI kg/m^2	27.5	5.9	21.0–34.8
HbA1c, %	7.2	0.5	6.5–7.8

3.2. Primary and Secondary Outcomes

The primary outcome PPG +iAUC$_{2hr}$ showed dose-related reductions from the addition of MFE (Table 2). The mean relative reduction of 22.4% for the 0.75 g dose of MFE was robust and exceeded the pre-specified target effect size of 15%. PPI tAUC$_{2hr}$ was also reduced in a dose-related way by the addition of MFE (Table 3), with a mean 17.5% reduction for the 0.75 g dose of MFE.

Table 2. Plasma glucose response over 2 hr following the consumption of mulberry fruit extract (MFE) added to boiled rice.

Intervention	N	Mean Glucose +iAUC$_{2hr}$ (Lower, Upper 95% CI), min·mmol/L	Mean % Difference, MFE vs. Control (Lower, Upper 95% CI)
Control	21	346 (307, 390)	
Control + 0.37 g MFE	22	318 (281, 360)	−8.2 (−20.8, 6.6)
Control + 0.75 g MFE	20	269 (215, 336)	−22.4 (−38.6, −1.9)

Table 3. Serum insulin response over 2 hr following the consumption of mulberry fruit extract (MFE) added to boiled rice.

Intervention	N	Mean Insulin tAUC$_{2hr}$ (Lower, Upper 95% CI), min·mIU/L	Mean % Difference, MFE vs. Control (Lower, Upper 95% CI)
Control	21	6470 (5464, 7661)	
Control + 0.37 g MFE	22	5847 (4949, 6908)	−9.6 (−20.7, 3.0)
Control + 0.75 g MFE	20	5334 (4501, 6325)	−17.5 (−27.9, −5.7)

There were minimal indications of any intolerance or safety issues, and none could be specifically attributed to the consumption of MFE. On the gastrointestinal discomfort questionnaire, one subject reported mild bloating at one visit and mild nausea at another, both at t = 240 min. One other subject reported mild bloating at baseline (t = −30 min) at one visit. No other symptoms of gastrointestinal discomfort were reported by any other subjects at any timepoints. In total, five adverse events were recorded for three subjects (one case each of vomiting, dizziness, and nausea and two cases of abdominal distension). All events were judged to be mild, did not require medical treatment, and were unrelated to the specific study products or procedures.

3.3. *Exploratory Outcomes*

The profiles of PPG and PPI responses over the full 4 hr postprandial period are shown in Supplementary Material, Figures S2 and S3. There were limited effects of MFE on the PPG responses when summed over 3 and 4 hr (+iAUC$_{3hr}$ and +iAUC$_{4hr}$), although these were slightly reduced by the higher MFE dose (Supplementary Material, Tables S2 and S3). However, both the 3 and 4 hr PPI responses (tAUC$_{3hr}$ and tAUC$_{4hr}$) were clearly reduced by the higher dose of MFE (Supplementary Material, Tables S4 and S5).

The addition of MFE produced modest, dose-related reductions in C_{max} (Supplementary Material, Table S6) and more consistent reductions in glucose swing (Supplementary Material, Table S7). Pooled urine glucose levels following the consumption of the test meals were highly variable but, on average, increased similarly from the baseline in all groups (Supplementary Material, Table S8).

4. Discussion

The addition of relatively low doses of MFE to rice led to dose-related reductions in the acute PPG and PPI responses in individuals with type 2 diabetes. This was most apparent during the immediate (2 hr) post-prandial period and was accompanied by apparent reductions in the post-meal peak glucose levels and glucose swing. The effects were more consistent and robust for a dose of 0.75 g of MFE than a 0.37 g dose, and at the higher dose, there were also reductions in PPI over the full 4 hr measurement timeframe.

The effects seen here were consistent with our previous research on individuals without diabetes, as was the absence of any indications of intolerance or other adverse effects of MFE. The mean reduction in PPG following the higher dose of MFE also exceeded our pre-defined 'desired' target reduction of $\geq$15%. Although we previously reported similar reductions in PPG and PPI with doses of 0.37 g of MFE containing ~1.85 mg of DNJ, that level appears to be close to the lower limit of efficacy [10]. A dose of 0.75 g of MFE may therefore be advised to ensure more reliable effects across different carbohydrate sources and populations [12].

DNJ in mulberry extracts has a well-known primary mechanism of action, similar to alpha-glucose-inhibiting drugs, which are effective in the treatment of diabetes and in reducing the risks of co-morbidities. The benchmark 15% change in PPG achieved here corresponds to about one-third of the observed effect on the PPG of drugs (miglitol, acarbose) used to reduce PPG [4]. This seems a reasonable effect size for a dietary adjunct, given that the drugs are available only with a prescription for patients under medical supervision and that gastrointestinal discomfort due to carbohydrate malabsorption is a common side-effect of these medications [26]. Some studies using high doses of MLE have reported increases in breath hydrogen, indicative of carbohydrate malabsorption [19,27]. Thaipitakwong et al. reported that an MLE dose containing 18 mg of DNJ resulted in a high incidence of bloating and flatulence, which was not seen at DNJ levels of 6 or 12 mg [28]. Comparisons between the MFE used here and studies using other mulberry extracts must be made with caution, however, due to differences in (often unreported) DNJ levels and overall product compositions.

We have found no indications of malabsorption or intolerance here or in studies using MFE at levels up to twice those in the present trial [10,13]. This may be due in part to the relatively low dose of DNJ and its high and rapid absorption in the proximal intestine, thus limiting the period that it is present in the gut lumen, affecting carbohydrate digestion [29]. Ingested DNJ is estimated to reach maximum plasma concentrations in about 30 min, much faster than common alpha-glucosidase-inhibiting drugs (acarbose, miglitol), which have a longer duration of presence and action in the gut [29,30]. The short period of activity of DNJ has implications for effect sizes and for its suitability for use with food, relative to medications. This is advantageous for mitigating potential side effects whilst still moderating the initial and peak glucose responses following meals, but it limits the effects over the longer (3–4 hr) post-prandial period, as observed here.

We and others mainly attribute the observed effects of mulberry extracts on PPG to the presence of DNJ. Although DNJ can be found in a limited number of other plants or can be synthesized by certain bacteria [29,31], to our knowledge, DNJ is not found in meaningful amounts in any dietary sources other than mulberries. It is, however, also possible that other iminosugars or minor components of MFE could also contribute to these effects.

While the present results are largely in line with our previous research on MFE and the wider literature on other mulberry extracts, there are a number of limitations to this study. Most mulberry extracts intended for food or supplement use are derived from processing the raw materials with water and ethanol [29]. Although the MFE used here is commercially available and standardized for its DNJ content, the exact production process is proprietary, and MFE derived from other sources may, therefore, differ in their content of other minor components. It is possible that some part of the variation between studies of mulberry extracts may be attributable to these differences in composition.

This was a fairly small pilot trial, not formally powered for inferential hypothesis testing. Although this was adequate for the current purpose, a larger test population would provide for more confident estimates of effect sizes. Venous blood was collected to limit the burden on subjects. This is unlikely to bias the results in relation to the effects of MFE but may produce lower and more variable glucose values than capillary blood [32]. Therefore, while this should not affect the general conclusions of this pilot trial, other methods of blood collection may allow for a more precise estimate of effect sizes.

Effects of MFE on PPI were a secondary objective, which was mainly to re-confirm that effects on PPG were not attributable to a disproportionate insulin response. The reductions in insulin responses are most likely an indirect result of reducing the rate of glucose uptake and thereby also reducing the stimulation of insulin secretion. However, possible effects on insulin secretion have not been directly tested. In future research, a greater focus on measures of insulin release and action may be advised for this population in particular, as well as considering longer and repeated exposures and markers of sustained glycemic control.

5. Conclusions

This trial adds to a series of studies on mulberry extracts and MFE specifically, supporting their safety and efficacy for reducing acute PPG and PPI responses to common dietary carbohydrate sources and different populations [10–12]. Together with the present results, the evidence indicates that a modest dose of 0.75 g of MFE containing 3.75 mg of DNJ consumed with a digestible carbohydrate source is likely to cause mean reductions in the range of about 10–25% in the 2 hr PPG AUC, as well as reduce PPI, peak glucose, and glucose swing in healthy individuals and individuals with diabetes.

Supplementary Materials: The following supporting information can be downloaded at: https://www.mdpi.com/article/10.3390/nu16142177/s1, Table S1: Inclusion and exclusion criteria; Table S2: Plasma glucose area under the curve response over 3 hr; Table S3: Plasma glucose area under the curve response over 4 hr; Table S4: Serum insulin area under the curve response over 3 hr; Table S5: Serum insulin area under the curve response over 4 hr; Table S6: Maximum glucose level (C_{max}) over 4 hr; Table S7: Glucose swing (C_{max}–C_{min}) over 4 hr; Table S8: Pooled urine glucose concentrations over 4 hr; Figure S1: Consolidated Standards of Reporting Trials [CONSORT] subject flow diagram; Figure S2: Plasma glucose response over 4 hr; Figure S3: Serum insulin response over 4 hr.

Author Contributions: Conceptualization, all authors; data curation and formal analysis, H.H.; methodology, all authors; project administration and supervision, T.K. and A.-R.H.; writing—original draft, D.J.M.; writing—review and editing, all authors. All authors have read and agreed to the published version of the manuscript.

Funding: This research was funded entirely by Unilever.

Institutional Review Board Statement: The trial was conducted in compliance with the Declaration of Helsinki, and the protocol and informed consent forms were approved as Protocol FDS-NAA-1633 by the Ashirwad Ethics Committee on 6 November 2014, the Medisys Clinisearch Ethical Review

Board on 17 January 2015, and the Therapeutic Drug Monitoring Laboratory Institutional Ethics Committee on 23 March 2015.

Informed Consent Statement: Informed consent was obtained from all subjects involved in the study.

Data Availability Statement: The datasets are not publicly available, as participants did not give express consent for this. Data described in the manuscript, code book, and analytic code can be made available upon request, pending application and approval.

Acknowledgments: The authors acknowledge the important support and contributions to this research from a number of current and former Unilever colleagues and the staff of the clinical testing sites and Lambda Therapeutics Research Ltd.

Conflicts of Interest: T.K. and A.-R.H. are employees of Unilever, the sponsor of the study. D.J.M., M.A., and H.H. were employees of Unilever at the time the research was carried out but have no current affiliation with the company. Unilever employees contributed to the analysis and interpretation of data and the writing of this report. Unilever employees had no part in the intervention study execution, participant contact, or data collection and were blind to treatment codes until these were revealed after the data were unlocked. The authors declare no other conflicts of interest related the topic of this research.

Abbreviations

C_{max}	Maximum recorded concentration of glucose in the post-prandial period
C_{min}	Minimum recorded concentration of glucose in the post-prandial period
DNJ	1-deoxynojirimycin
iAUC	Incremental area under the curve
MFE	Mulberry fruit extract
MLE	Mulberry leaf extract
PPG	Postprandial glucose
PPI	Postprandial insulin
tAUC	Total area under the curve

References

1. Ceriello, A.; Colagiuri, S. International Diabetes Federation guideline for management of postmeal glucose: A review of recommendations. *Diabet. Med.* **2008**, *25*, 1151–1156. [CrossRef] [PubMed]
2. American Diabetes Association. 6. Glycemic targets: Standards of Medical Care in Diabetes. *Diabetes Care* **2019**, *42*, S61–S70. [CrossRef] [PubMed]
3. Hershon, K.S.; Hirsch, B.R.; Odugbesan, O. Importance of postprandial glucose in relation to A1C and cardiovascular disease. *Clin. Diabetes* **2019**, *37*, 250–259. [CrossRef] [PubMed]
4. Alssema, M.; Ruijgrok, C.; Blaak, E.E.; Egli, L.; Dussort, P.; Vinoy, S.; Dekker, J.M.; Denise Robertson, M. Effects of alpha-glucosidase-inhibiting drugs on acute postprandial glucose and insulin responses: A systematic review and meta-analysis. *Nutr. Diabetes* **2021**, *11*, 11. [CrossRef]
5. van de Laar, F.A.; Lucassen, P.L.B.J.; Akkermans, R.P.; van de Lisdonk, E.H.; Rutten, G.E.H.M.; van Weel, C. Alpha-glucosidase inhibitors for type 2 diabetes mellitus. *Cochrane Database Syst. Rev.* **2005**, *2005*, CD003639. [CrossRef] [PubMed]
6. Di Stefano, E.; Oliviero, T.; Udenigwe, C.C. Functional significance and structure–activity relationship of food-derived α-glucosidase inhibitors. *Curr. Opin. Food Sci.* **2018**, *20*, 7–12. [CrossRef]
7. Thakur, K.; Zhang, Y.-Y.; Mocan, A.; Zhang, F.; Zhang, J.-G.; Wei, Z.-J. 1-Deoxynojirimycin, its potential for management of non-communicable metabolic diseases. *Trends Food Sci. Technol.* **2019**, *89*, 88–99. [CrossRef]
8. Tseng, P.S.; Ande, C.; Moremen, K.W.; Crich, D. Influence of side chain conformation on the activity of glycosidase inhibitors. *Angew. Chem.* **2023**, *135*, e202217809. [CrossRef]
9. Compain, P.; Martin, O.R. *Iminosugars: From Synthesis to Therapeutic Applications*; John Wiley & Sons: Chichester, UK, 2007.
10. Mela, D.J.; Cao, X.-Z.; Govindaiah, S.; Hiemstra, H.; Kalathil, R.; Lin, L.; Manoj, J.; Mi, T.; Verhoeven, C.H. Dose-response efficacy of mulberry fruit extract for reducing post-prandial blood glucose and insulin responses: Randomized trial evidence in healthy adults. *Br. J. Nutr.* **2023**, *129*, 771–778. [CrossRef]
11. Boers, H.M.; van Dijk, T.H.; Duchateau, G.S.; Mela, D.J.; Hiemstra, H.; Hoogenraad, A.-R.; Priebe, M.G. Effect of mulberry fruit extract on glucose fluxes after a wheat porridge meal: A dual isotope study in healthy human subjects. *Eur. J. Clin. Nutr.* **2023**, *77*, 741–747. [CrossRef]
12. Mela, D.J.; Boers, H.M.; Kadam, T.; Hiemstra, H.; Kalathil, R.; Seijen ten Hoorn, J.W. Effect of mulberry fruit extract on post-prandial glycaemic and insulinemic responses to different rice types: A randomised trial in healthy adults. *Br. J. Nutr.* **2023**, *130*, 2088–2094. [CrossRef] [PubMed]

13. Mela, D.J.; Cao, X.-Z.; Dobriyal, R.; Fowler, M.I.; Li, L.; Manoj, J.; Mulder, T.J.P.; Murray, P.; Peters, H.P.F.; Vermeer, M.A.; et al. The effect of 8 plant extracts and combinations on post-prandial blood glucose and insulin responses in healthy adults: A randomized controlled trial. *Nutr. Metab.* **2020**, *17*, 51. [CrossRef]
14. Phimarn, W.; Wichaiyo, K.; Silpsavikul, K.; Sungthong, B.; Saramunee, K. A meta-analysis of efficacy of Morus alba Linn. to improve blood glucose and lipid profile. *Eur. J. Nutr.* **2017**, *56*, 1509–1521. [CrossRef] [PubMed]
15. Jeong, H.I.; Jang, S.; Kim, K.H. *Morus alba* L. for Blood Sugar Management: A Systematic Review and Meta-Analysis. *Evid. Based Complement. Altern. Med.* **2022**, *2022*, 9282154. [CrossRef]
16. Chen, X.; Sohouli, M.H.; Nateghi, M.; Melekoglu, E.; Fatahi, S. Impact of mulberry consumption on cardiometabolic risk factors: A systematic review and meta-analysis of randomized-controlled trials. *J. Clin. Pharm. Ther.* **2022**, *47*, 1982–1993. [CrossRef] [PubMed]
17. Asai, A.; Nakagawa, K.; Higuchi, O.; Kimura, T.; Kojima, Y.; Kariya, J.; Miyazawa, T.; Oikawa, S. Effect of mulberry leaf extract with enriched 1-deoxynojirimycin content on postprandial glycemic control in subjects with impaired glucose metabolism. *J. Diabetes Investig.* **2011**, *2*, 318–323. [CrossRef] [PubMed]
18. Hwang, S.H.; Li, H.M.; Lim, S.S.; Wang, Z.; Hong, J.-S.; Huang, B. Evaluation of a standardized extract from Morus alba against α-glucosidase inhibitory effect and postprandial antihyperglycemic in patients with impaired glucose tolerance: A randomized double-blind clinical trial. *Evid. Based Complement. Altern. Med.* **2016**, *2016*, 8983232. [CrossRef]
19. Mudra, M.; Ercan-Fang, N.; Zhong, L.; Furne, J.; Levitt, M. Influence of Mulberry Leaf Extract on the Blood Glucose and Breath Hydrogen Response to Ingestion of 75 g Sucrose by Type 2 Diabetic and Control Subjects. *Diabetes Care* **2007**, *30*, 1272–1274. [CrossRef]
20. Nakamura, S.; Hashiguchi, M.; Yamaguchi, Y.; Oku, T. Hypoglycemic effects of Morus alba leaf extract on postprandial glucose and insulin levels in patients with type 2 diabetes treated with sulfonylurea hypoglycemic agents. *J. Diabetes Metab.* **2011**, *10*, 2155–6156. [CrossRef]
21. Riche, D.M.; Riche, K.D.; East, H.E.; Barrett, E.K.; May, W.L. Impact of mulberry leaf extract on type 2 diabetes (Mul-DM): A randomized, placebo-controlled pilot study. *Complement. Ther. Med.* **2017**, *32*, 105–108. [CrossRef]
22. Kim, J.Y.; Ok, H.M.; Kim, J.; Park, S.W.; Kwon, S.W.; Kwon, O. Mulberry leaf extract improves postprandial glucose response in prediabetic subjects: A randomized, double-blind placebo-controlled trial. *J. Med. Food* **2015**, *18*, 306–313. [CrossRef] [PubMed]
23. Julious, S.A. Sample size of 12 per group rule of thumb for a pilot study. *Pharm. Stat.* **2005**, *4*, 287–291. [CrossRef]
24. Thomas, D.; Elliott, E. The use of low-glycaemic index diets in diabetes control. *Br. J. Nutr.* **2010**, *104*, 797–802. [CrossRef] [PubMed]
25. Greenwood, D.C.; Threapleton, D.E.; Evans, C.E.; Cleghorn, C.L.; Nykjaer, C.; Woodhead, C.; Burley, V.J. Glycemic index, glycemic load, carbohydrates, and type 2 diabetes: Systematic review and dose–response meta-analysis of prospective studies. *Diabetes Care* **2013**, *36*, 4166–4171. [CrossRef]
26. Hedrington, M.S.; Davis, S.N. Considerations when using alpha-glucosidase inhibitors in the treatment of type 2 diabetes. *Expert Opin. Pharmacother.* **2019**, *20*, 2229–2235. [CrossRef] [PubMed]
27. Nakamura, M.; Nakamura, S.; Oku, T. Suppressive response of confections containing the extractive from leaves of Morus Alba on postprandial blood glucose and insulin in healthy human subjects. *Nutr. Metab.* **2009**, *6*, 29. [CrossRef] [PubMed]
28. Thaipitakwong, T.; Supasyndh, O.; Rasmi, Y.; Aramwit, P. A randomized controlled study of dose-finding, efficacy, and safety of mulberry leaves on glycemic profiles in obese persons with borderline diabetes. *Complement. Ther. Med.* **2020**, *49*, 102292. [CrossRef] [PubMed]
29. Parida, I.S.; Takasu, S.; Nakagawa, K. A comprehensive review on the production, pharmacokinetics and health benefits of mulberry leaf iminosugars: Main focus on 1-deoxynojirimycin, d-fagomine, and 2-O-α-d-galactopyranosyl-DNJ. *Crit. Rev. Food Sci. Nutr.* **2023**, *63*, 3468–3496. [CrossRef] [PubMed]
30. Gao, K.; Zheng, C.; Wang, T.; Zhao, H.; Wang, J.; Wang, Z.; Zhai, X.; Jia, Z.; Chen, J.; Zhou, Y.; et al. 1-Deoxynojirimycin: Occurrence, extraction, chemistry, oral pharmacokinetics, biological activities and in silico target fishing. *Molecules* **2016**, *21*, 1600. [CrossRef]
31. Zhang, W.; Mu, W.; Wu, H.; Liang, Z. An overview of the biological production of 1-deoxynojirimycin: Current status and future perspective. *Appl. Microbiol. Biotechnol.* **2019**, *103*, 9335–9344. [CrossRef]
32. Brouns, F.; Bjorck, I.; Frayn, K.; Gibbs, A.; Lang, V.; Slama, G.; Wolever, T. Glycaemic index methodology. *Nutr. Res. Rev.* **2005**, *18*, 145–171. [CrossRef] [PubMed]

nutrients

MDPI

Article

The Study of Myo-Inositol's Anxiolytic Activity on Zebrafish (*Danio rerio*)

Maria Derkaczew [1,2,*], Bartosz Kędziora [1,2], Małgorzata Potoczna [3], Piotr Podlasz [3], Krzysztof Wąsowicz [3], Marcin Jóźwik [4,*] and Joanna Wojtkiewicz [1,*]

1 Department of Human Physiology and Pathophysiology, School of Medicine, Collegium Medicum, University of Warmia and Mazury, 10-082 Olsztyn, Poland

2 Students' Scientific Club of Pathophysiologists, Department of Human Physiology and Pathophysiology, School of Medicine, University of Warmia and Mazury, 10-082 Olsztyn, Poland

3 Department of Pathophysiology, Forensic Veterinary Medicine and Administration, Faculty of Veterinary Medicine, University of Warmia and Mazury, 10-719 Olsztyn, Poland; malgorzata.mierzejewska@transpharmation.co.uk (M.P.); krzysztof.wasowicz@uwm.edu.pl (K.W.)

4 Department of Gynecology and Obstetrics, Collegium Medicum, University of Warmia and Mazury, 10-045 Olsztyn, Poland

* Correspondence: m.derkaczew@gmail.com (M.D.); marcin.jozwik@uwm.edu.pl (M.J.); joanna.wojtkiewicz@uwm.edu.pl (J.W.)

Abstract: Introduction: Myo-inositol (MI) is the most abundant inositol found in nature. To date MI supplementation is reported to be effective in the treatment of polycystic ovary syndrome, it is also suggested to alleviate the symptoms of diabetes and neurodegenerative disorders, but to date no statistically significant effects of inositol on depressive and anxiety symptoms were proven. In the study of anxiolytic effects in zebrafish, we often use the thigmotaxis index measuring the ratio of the amount of time the animal spends near the walls compared to the entire arena. Aim: The objective of this paper was to examine the effect of MI on zebrafish embryos' locomotor activity, as well as its potential anxiolytic activity in zebrafish larvae. Material and methods: In the first part of the experiment, the embryos were incubated with 5, 10, 20, and 40 mg/mL MI. 1-day post fertilization, embryo mobility was evaluated and burst activity was calculated. In the next part of the study, the behavior of 5-day-old larvae was tested. Results: Tests on embryo movement showed an increase in burst activity in the MI group at concentrations of 40 mg/mL ($p < 0.0001$) and a slight decrease in the group at concentrations of 10 mg/mL ($p < 0.05$). MI in the light/dark challenge had no impact on the thigmotaxis index. Conclusions: MI was shown to not affect stress reduction in zebrafish larvae. Further research on the potential of MI and other stereoisomers is needed.

Keywords: myo-Inositol; anxiety; anxiolytics; zebrafish

1. Introduction

Herbal medicine and a healthy lifestyle based on a balanced diet full of fruits, vegetables, nuts and seeds have been promoted for years. Cyclitols are a group of compounds that may have a beneficial effect on human health and support immunity against numerous diseases, but its use as a clinical therapeutic is not established. Also known as sugar alcohols or polyols, they are widely distributed in the environment, and are found in many organic products such as citrus fruits, nuts, yeast, grains, beans, buckwheat, and many others [1]. Cyclitols and their derivatives perform many functions in eukaryotic cells. They regulate ion channel activity, intracellular phosphate storage, cell wall formation, signal transduction, membrane biogenesis and osmoregulation; furthermore, they are the forerunners of crucial secondary messengers [2]. Myo-inositol (MI) is a representative of the cyclitol group, at the same time being the most abundant inositol in nature. MI was first isolated from muscle extracts in 1850 by Scherer and since then has been intensively studied for its possible beneficial effects on the human body [3]. MI and its derivative supplementation are characterized by good drug tolerance, low toxicity, and possible use by pregnant women [4]. Figure 1 is depicting the chemical structures of MI and its derivatives with possible anxiolytic effects.

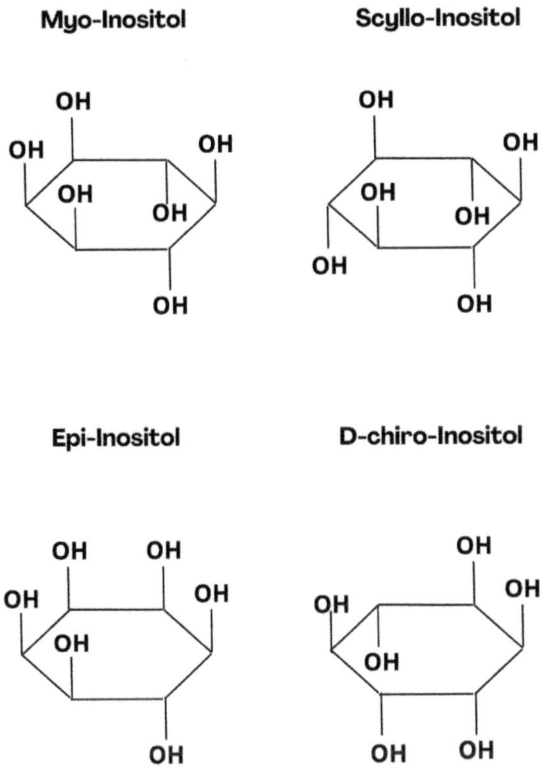

Figure 1. Chemical structure of Myo-Inositol and its derivatives with possible anxiolytic effect.

Since the start of the COVID-19 pandemic, the number of people suffering from depression and anxiety has increased substantially: a 27.6% increase in cases of major depressive disorders and a 25.6% increase in anxiety disorder cases [5]. Researchers have emphasized that it was a predictable phenomenon in the face of an unknown emerging crisis and that these numbers should normalize over time [6,7]. It is clear that during the pandemic onset, there was an acute increase in mental health symptoms [6]. The exact impact of the pandemic on the human mental condition needs to be determined by further longitudinal study [8]. The use of nutrient-based "nutraceuticals" and plant-based "phytoceuticals" to treat mental disorders is widespread. Still, no updated global clinical guidelines have been issued since 2015, until a recent meta-analysis by Sarris et al. where inositol was not advised for treating depression and lacked sufficient evidence for its effectiveness in treating anxiety [9]. Anxiety is one of the most widespread psychiatric disorders and it is still a significant problem among many patients suffering from other somatic diseases [10,11]. Currently used psychiatric drugs have many side effects and can lead to addiction. Natural methods of combating depression and drugs that would additionally carry a lower risk of complications are still being sought [12].

The best animal models for testing possible new treatment ideas are still being sought. Mouse models are expensive and require large quantities of samples, which is why an alternative is needed. For these reasons, the zebrafish (*Danio rerio*) has recently become the prominent model organism for research in many fields, such as toxicology, drug discovery, developmental biology, oncology, molecular genetics, and many others [13]. In the 1970s, George Streisinger from the University of Oregon was the first one to use the zebrafish as a model organism due to its low cost, the possibility of easier genetic manipulation than in mice models, and a fast development cycle. Zebrafish have many similarities in physiology and genetics to humans; moreover, 70% of disease genes function the same in humans and zebrafish [13]. Zebrafish enable high throughput screening and are compatible with the standard multi-well plates and video-recording systems used in industry [14].

The larvae of zebrafish are exceptionally applicable for behavioral tests performed in multi-well plates due to their small size and low-cost maintenance. Furthermore, these models are ideal for high throughput screening because of their relative maturity compared to adults in terms of swimming ability, motor functionality, sensitivity to stress, and their ability to perform simple motor tasks while receiving appropriate signals from the environment [14]. While the nervous system in zebrafish is less complex than in Homo sapiens, the action of the relevant neurotransmitters and adequate changes in their behavior in response to stress can still be observed. Researchers have used many tests to investigate zebrafish anxiety levels, such as the novel tank dive, open field test, light-dark test, startle test, electric shock assay, and others [15,16]. Thigmotaxis, also known as wall-hugging, is the tendency of the zebrafish to avoid the center of the arena and choose to stay near the edge of the well. Also, changes in the intensity of light affect the level of anxiety in zebrafish with the tendency to avoid the darkness. Such a behavior is commonly seen in nature in many species and demonstrates a peculiar reaction to stress [14,17].

The objective of our work was to examine the effect of MI on zebrafish embryos' locomotor activity, as well as its potential anxiolytic activity in zebrafish larvae.

2. Materials and Methods

2.1. The Fish Maintenance and Ethic Statement

All fish lines are housed in the fish facility of the Laboratory of Genomics and Transcriptomics, University of Warmia and Mazury in Olsztyn, Poland—built according to the local animal welfare standards. According to the European Directive 2010/63/EU and Polish legal regulations O.J. of 2015, item 266, studies performed on early-life-stage zebrafish larvae and euthanasia do not require Ethic Committee permissions.

2.2. Zebrafish Spawning, Embryo Selection, and Larvae Incubation

The adult AB zebrafish strain was set for spawning, with a female-to-male ratio of approximately 1:1, in spawning containers. The spawning was induced by turning on the light in the morning. Eggs in the stadium of 1 h post fertilization were collected and washed with E3 medium (5 mM NaCl, 0.17 mM KCL, 0.33 mM $CaCl_2$-H_2O, 0.33 mM $MgCl_2$-$6H_2O$, and pH 7.2). Collected embryos were selected and placed randomly in 5 Petri dishes with prepared solutions of MI (Chemat, Gdańsk, Poland). Four experimental groups were done consecutively: 5 mg/mL, 10 mg/mL, 20 mg/mL, and 40 mg/mL of MI diluted in E3 solution. The control group was incubated in the E3 medium. Prepared plates were incubated at 28.5 °C for 24 h.

2.3. Embryo Movement Analysis

Tail coiling analysis was conducted on healthy 24 hpf embryos. Embryos in each treatment were selected to take 5-min videos under the stereomicroscope (SteREO Discovery.V8, Zeiss, Germany). Analysis was performed using DanioScope Software v 1.2.206 (Noldus, Wageningen, The Netherlands) [18]. Each group of embryos was recorded for 5 min. DanioScope automatedly analyses the video recordings of larvae and evaluates the parameters. Tail coiling activity was expressed as the proportion of the time of flicked tail in 1 min [%]. The proportion of the time of spontaneous head-tail contraction–burst activity [%] and total burst duration [s] of each embryo was counted and the average for each group was calculated.

2.4. Larvae Behavioural Assessment

All groups were incubated for 120 h. After every 24 h, solutions in all culture plates were changed. At 120 hpf, 24 larvae from each tested group were randomly chosen and transferred to 24-well plates (one individual larva per well) for behavioural tests.

The measurements were performed using the DanioVision system with Ethovision XT v.15 software (Noldus, Wageningen, The Netherlands) [19,20]. The observation time for each larva was 25 min, of which the first 5 min were given for adaptation. The larvae spent the next 10 min in the light, and the last 10 min were spent in the dark. The distance moved [mm] was calculated for each larva.

To assess thigmotaxis, the swimming arena must provide adequate space for distinguishing between inner and outer zones. We used the 24-well plate format (diameter 16.2 mm). Figure 2 is a depiction of the schematic representation of how the inner and outer zones were defined.

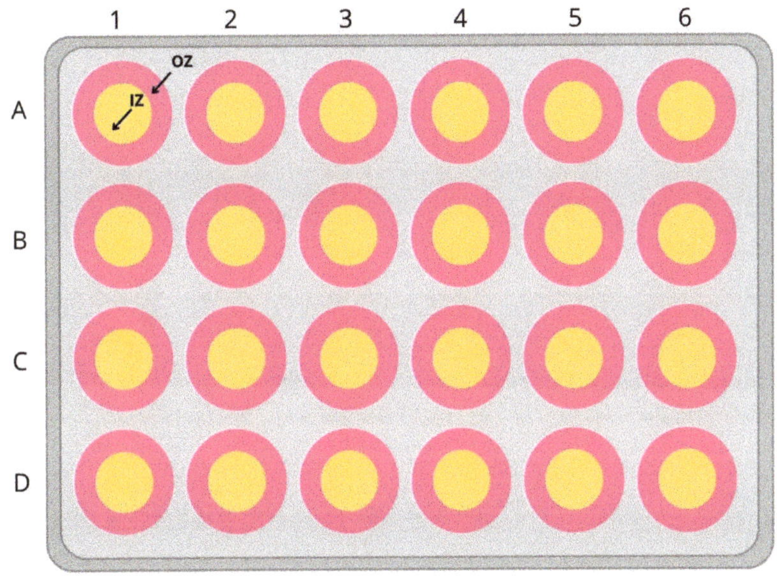

Figure 2. Schematic illustration of 24-well plate for larvae incubation (IZ—inner zone, OZ—outer zone).

The thigmotaxis was calculated as the ratio between total distance moved (TDM) in the outer zone of the test apparatus and TDM over the entire test arena (including inner and outer zones). The % of TDM in the outer zone was obtained by multiplying this ratio by 100 [14].

2.5. Statistical Analysis

All data were identified as normally or non-normally distributed using the Shapiro-Wilk test and expressed as a mean with standard error of the mean (SEM) to display more representative data. The quality of the two populations was tested using the Mann–Whitney U test. Multiple comparisons for comparing two or more independent samples between groups were performed using the Kruskal-Wallis test with Bonferroni correction. The data used to analyze the thigmotaxis index were cleaned by manually removing the extreme values (0% and 100%) and then removing outliers using the Grubbs test. $p < 0.05$ indicated statistical significance (ns: $p > 0.05$, *: $p < 0.05$, **: $p < 0.01$, ***: $p < 0.001$, ****: $p < 0.0001$). Statistical analyses were performed using R software (version 4.3.3) and the 'ggstatsplot' and 'ggplot2' approach.

3. Results

3.1. *The Movement Analysis of Embryos*

The Mann-Whitney U test showed an increase in burst activity in the control group and MI at concentrations of 40 mg/mL ($p < 0.0001$), and a slight decrease at concentrations of 10 mg/mL ($p < 0.05$). During the total burst duration analysis, a significant increase was observed when the larvae were treated with MI at a concentration of 40 mg/mL ($p < 0.0001$) compared to the E3-treated group. Additionally, a slight decrease in total burst duration was observed with MI at 10 mg/mL ($p < 0.05$). The results of zebrafish embryo locomotor activity are shown in Figure 3.

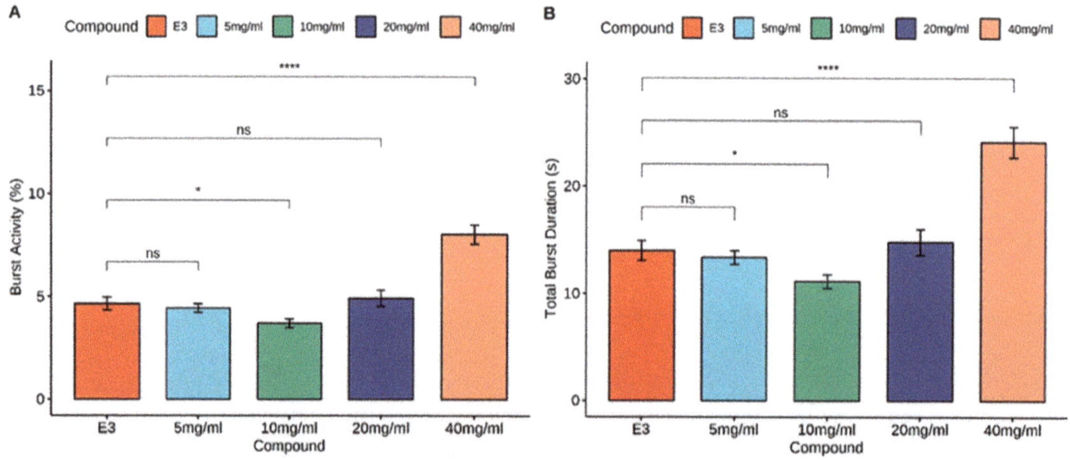

Figure 3. Effects of myoinositol (5, 10, 20, and 40 mg/mL) and E3 on locomotor activity. (**A**) Burst Activity. (**B**) Total Burst Duration. Data are presented as the mean ± SEM; $n_{E3} = 39$, $n_{5\ mg/mL} = 37$, $n_{10\ mg/mL} = 37$, $n_{20\ mg/mL} = 35$, $n_{40\ mg/mL} = 37$. ns > 0.05, * $p < 0.05$, **** $p < 0.0001$ n comparison to control group; (U Mann-Whitney test).

3.2. *Larvae Locomotor Analysis Using Dark-Light Test*

3.2.1. Larvae Locomotor Activity

Figure 4 shows the results of locomotor activity testing under alternating light and dark conditions. Larvae were exposed to myo-inositol (5, 10, 20, and 40 mg/mL) or vehicle (E3 solution) for 25 min. Testing began in the adaptive stage (5 min), followed by one cycle of light (10 min) and darkness (10 min). Locomotion was not analyzed in the first stage.

Data in Figure 4 are presented as the mean distance moved (in mm) and SEM in 1-min intervals during the 25-min sessions. For every group, the activity increased during the initial dark period (shaded area).

The U Mann-Whitney test showed an increase in locomotor activity during the dark phase in the control group and MI at concentrations of 5 mg/mL, 10 mg/mL, 20 mg/mL, and 40 mg/mL ($p < 0.0001$) compared to the light phase. The Kruskal-Wallis test revealed statistically significant changes in zebrafish larvae behavior after incubation in myoinositol solutions in light [$\chi^2(4) = 38.12$, $p < 0.0001$] and dark [$\chi^2(4) = 39.88$, $p < 0.0001$] conditions. The post hoc Bonferroni's test demonstrated a decrease in locomotor activity during the light phase in myoinositol at concentrations of 5 mg/mL ($p < 0.05$), 20 mg/mL ($p < 0.01$), and 40 mg/mL ($p < 0.0001$) compared to the control group (E3). During the dark phase of the experiment, a significant decrease in locomotor activity was observed after incubation in myoinositol at concentrations of 5 mg/mL ($p < 0.05$) and 40 mg/mL ($p < 0.0001$) compared to the E3 control group in the dark phase. The effects of MI and E3 on locomotor activity are shown in Figure 5.

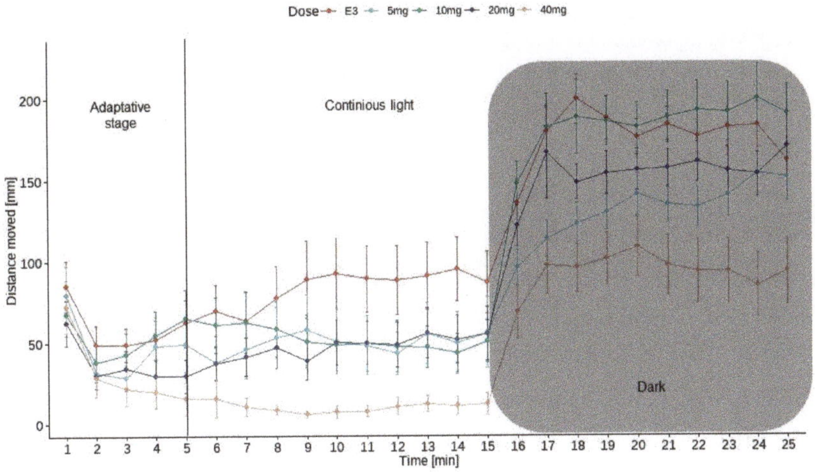

Figure 4. Light/dark challenge with larvae fish. Locomotor activity was recorded for 15 min before a sudden dark period lasting 10 min (gray shadowing). The fish were recorded for the entire 25 min; $n = 24$.

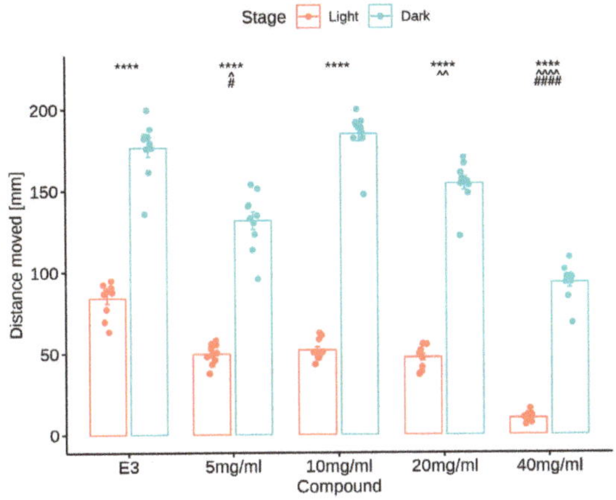

Figure 5. Effects of myoinositol (5, 10, 20, and 40 mg/mL) and E3 on locomotor activity. Average distance moved by zebrafish larvae within each 1-min time bin under either light (red bars) or dark (blue bars) conditions. Data are presented as mean ± SEM; n = 20. **** $p < 0.0001$ in comparison to light conditions within the same concentration group (U Mann-Whitney test); ^ $p < 0.05$, ^^ $p < 0.01$, ^^^^ $p < 0.0001$ in comparison to the control group under light conditions; # $p < 0.05$, #### $p < 0.0001$ in comparison to the control group under dark conditions (post hoc Bonferroni's test).

3.2.2. Thigmotaxis Index

Myoinositol treatment in the light/dark challenge had no impact on the thigmotactic behaviors of larvae (Kruskal-Wallis test: treatment under dark conditions [$\chi^2(4) = 4.59$, $p = 0.33$] nor was there an impact under light conditions ($\chi^2(4) = 1.35$, $p = 0.85$). The Mann-Whitney U test showed no change in the % TDM in the outer zone by zebrafish larvae after E3 ($p > 0.05$) and myoinositol treatments at concentrations of 5, 10, 20, and 40 mg/mL ($p > 0.05$) in the light phase of the experiment compared to the dark phase. During the dark challenge phase, no significant changes in the % TDM in the outer zone were observed when the larvae were treated with myoinositol at any concentration compared with the E3-treated group in the dark phase ($p > 0.05$). During the light phase of the experiment, no significant changes in the % TDM in the outer zone were observed when larvae were treated with myoinositol at any concentration, compared to the E3-treated group in the light phase ($p > 0.05$). The effects of MI on the thigmotaxis index are shown in Figure 6.

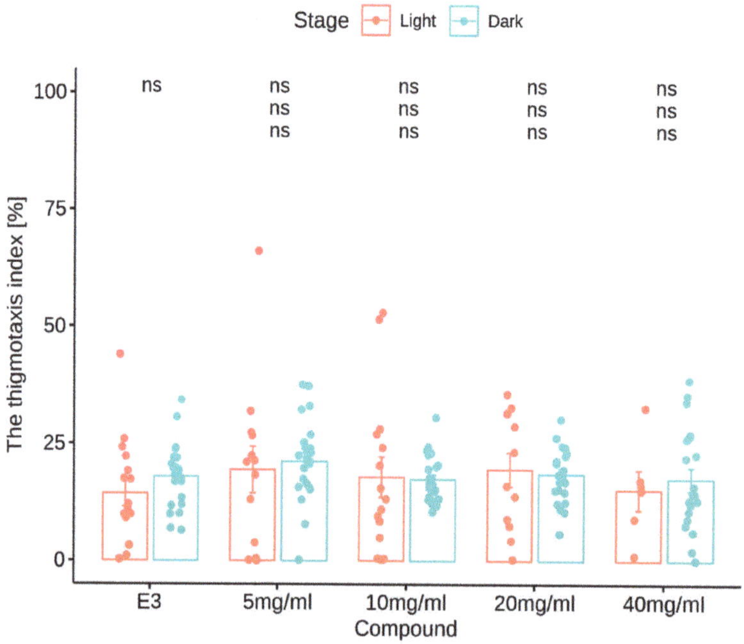

Figure 6. Effects of myoinositol (5, 10, 20, and 40 mg/mL) and E3 on thigmotaxis index measured as the % TDM in the outer zone. The thigmotaxis index by zebrafish larvae within each 1-min time bin under either light (blue bars) or dark (red bars) conditions. Data are presented as the mean ± SEM; ns (first upper row) $p > 0.05$ in comparison to light conditions within the same concentration group (U Mann-Whitney test); ns (second row) $p < 0.05$ in comparison to the control group under dark conditions; ns (third row) > 0.05, in comparison to the control group under light conditions (post hoc Bonferroni's test).

4. Discussion

Despite many years of research on the effects of MI, there are still no clear results of its use, and further research is needed to determine its effectiveness. It has already been speculated that supplementation with MI alleviates the insulin resistance symptoms in diabetic and Polycystic Ovary Syndrome (PCOS) patients, but the results of studies found in the literature are not clear. In some sources, we can find information about the MI's ability to control insulin activity, improve insulin sensitivity, and has salutary effects in people with diabetes. Sharma et al. report that MI has the potential to regulate insulin, aiding in the prevention and management of diabetes mellitus [21]. Additionally, it has the advantage of being an inexpensive and safe alternative to commonly used drugs [22,23]. Unfer et al. performed an analysis of various studies on MI supplementation for improving hormonal imbalances in PCOS and provide the level Ia evidence of MI's effectiveness [24]. In a study by Zhao et al. MI multicomponent supplementation with D-chiro-Inositol (DCI) was ranked best at improving menstrual frequency and this combination was found to be superior the supplementation of MI or DCI alone [25]. However, there are still few studies on the anxiolytic and antidepressant effects of MI. There are reports of reduced levels of brain MI in patients with depression and anxiety compared to healthy controls and low plasma MI concentrations have been suggested as a possible marker of major depressive disorders [26–29]. Also, PCOS patients often suffer from depression or anxiety, related to hormonal disturbances and weight gain. Cantelmi et al. report that MI supplementation turned out to reduce these symptoms [30]. In studies concerning possible antidepressant and anxiolytic inositol activity, researchers used MI and its stereoisomers such as Epi-Inositol (EI), Scyllo-Inositol (SI), and also D-pinitol; an analog of MI widely found in many plant families [30–32]. Most of these studies suggest the positive effect of MI supplementation on depression and anxiety [33]. However, as we mentioned before the recent guidelines by the World Federation of Societies of Biological Psychiatry (WFSBP) report that inositol is not advised for treating depression and lacks sufficient evidence for its effectiveness in treating anxiety Mashayekh-Amiri et al. [25]. Ref. [34] report an improvement in sleep quality in pregnant women supplemented with MI. SI contributes to the reduction of symptoms of depression and anxiety in patients with Alzheimer's Disease [35]. EI is a stereoisomer absent in mammalian tissue, but present in pine bark. In research from Einat et al., the plus-maze model of anxiety in mice was used to test the anxiolytic-like activity of EI and MI. EI had the strongest anxiolytic-like effect, followed by MI, in comparison to the control group [36–38]. D-pinitol has also been shown to have an antidepressant and anxiolytic effect [39]. In a review of the plant-based methods of anxiety and depression treatment by Fajemiroye et al., researchers list a plant that originated in Brazil, *Mimosa pudica*, in which D-pinitol is one of the active principles. It has an antidepressant-like effect, mediated by the serotonergic system. It is used in the treatment of depression and insomnia and is consumed in the form of an infusion of dried leaves [40]. Alonso-Castro et al. analyzed the effects of an ethanol extract of *Senna septemtrionalis* aerial parts (consisting of 42% D-pinitol) on mice behavior. D-pinitol showed anxiolytic-like activity in the four anxiety models and this effect turned out to be stronger than after fluoxetine administration [39].

This is the second research to date in which MI activity was tested on the *Danio rerio*. In this study, we used concentrations of MI with a maximum value of 40 mg/mL. Based on our previous research, in which we conducted the fish embryo toxicity (FET) test, MI concentrations higher than 40 mg/mL turned out to be toxic and led to significant disturbances in zebrafish maturation and survivability [1]. In another research paper found in online databases, the authors investigated the impact of MI on stress reduction in another fish species—the turbot (*Scophthalmus maximus*). The researchers concluded that MI significantly reduces stress in individuals in response to salinity. MI was administered by dissolving it in the aquatic environment where the individuals lived and by adding it to their feed. The genetic testing of the transcriptome of the experimental group showed that MI increased the effectiveness of physiological processes such as steroid biosynthesis, steroid metabolism, circadian rhythm, tryptophan metabolism, metabolism of xenobiotics by cytochrome P450, oxidoreductase activity, iron ion binding, and heme binding [41].

The tail-coiling activity measurement performed via video recording of embryos is a relatively new alternative method for screening developmental neurotoxicity induced by tested substances. Tail coiling activity can be assessed based on many factors and one of them is burst activity; the percentage of time the embryo is moving in the recorded time bin [42,43]. In our work, burst activity of embryos was calculated and a significant increase in burst duration was observed at the highest dose of MI (40 mg/mL), but a minimal decrease in burst activity was detected in the 10 mg/mL group.

In this paper, we focused on determining the possible MI's anxiolytic activity and behavioral changes in Zebrafish embryos and larvae. In the behavioral assessment, researchers often use the visual motor response (VMR) test which is based on a sudden transition in the illumination of the observed fish from light to darkness. Sudden cut-off of the light source causes a significant robust increase in the zebrafish locomotor activity due to the stress response [14,44]. In our study for every tested group, the activity increased during the initial part of the dark period, which is in line with other authors' outcomes. Zebrafish larvae are showing aversion toward darkness, which leads to an increase of the traveled distance when compared to the light phase. In our research, average distance moved was higher in the dark than in the light conditions in every group.

In the analysis of zebrafish larvae behavior, specific endpoints are used, such as total distance traveled, velocity, and time spent in zones, which allows for statistical data comparison [45]. In the papers concerning the anxiolytics tested on zebrafish, the total distance traveled by fish in combination with dark and light periods allows the assessment of anxiolytic or sedative effects. Zebrafish larvae show natural darkness avoidance (i.e., scotophobia) [46]. The results of our study remain ambiguous. While the characteristic zebrafish behaviors such as scotophobia remained unchanged, indicators showing changes in levels of anxiety were not significantly affected by MI. MI was found to reduce the distance moved in comparison to the control group. The main limitation of our study is that it is based on simple behavioral analysis. Our future studies will require biochemical validation as well as tracking the transcriptome changes in zebrafish.

Thigmotaxis is a validated index for the evaluation of anxiety changes in animals and humans. Thigmotaxis refers to the inclination to steer clear of the central area of an arena and instead remain close to or move along the boundaries of the surroundings [14,47]. The thigmotaxis index in experimental groups showed no significant changes than in the control group. Therefore, a significant anxiolytic effect of MI on zebrafish could not be demonstrated.

5. Conclusions

Plant-based substances that can be used as potential anxiolytics are still being sought due to the many side effects of currently available drugs. In this project, we analyzed the effects of MI on zebrafish behavior, development, and potential anxiolytic effects. No direct anxiolytic effect of MI on zebrafish was demonstrated, but MI boosted the 24-hpf embryo locomotor activity and therefore it may have a beneficial role in the development of zebrafish embryos. It is worth noting that the group of cyclitols includes numerous derivatives, and dietary supplements containing MI are commonly used daily with remarkable effects. Still there are a few clinical trials registered, complited and with published results—20 studies to date about the effects of the whole inositol group, and none concerning the topic of depressive and anxiety disorders. Our research paves the way for further research on the action and effectiveness of MI and its other stereoisomers. Further studies are needed for a deeper understanding of the anxiolytic and developmental effects of MI, including transcriptome analysis using next-generation sequencing.

Author Contributions: Conceptualization, M.D. and J.W.; methodology, M.D., M.P., M.J., K.W. and P.P.; validation, M.D., B.K., M.P., P.P. and J.W.; formal analysis, M.D.; investigation, M.D., B.K., M.P., P.P. and J.W.; data curation, M.D.; writing—original draft preparation, M.D.; writing—review and editing, M.D., B.K., M.P., P.P. and J.W.; visualization, M.D., B.K. and M.P.; supervision, J.W., K.W., M.J. and P.P.; project administration, J.W., M.J., K.W. and P.P. All authors have read and agreed to the published version of the manuscript.

Funding: This research received no external funding.

Institutional Review Board Statement: All fish were housed in the fish facility of the Department of Pharmacology and Toxicology, School of Medicine Laboratory, University of Warmia and Mazury in Olsztyn, Poland, which was built according to the local animal welfare standards. All animal procedures were performed in accordance with Polish and European Union animal welfare guidelines. According to the European Directive 2010/63/EU and Polish law regulations, O.J. of 2015, item 266, all procedures performed in the present study including the use of early life-stage zebrafish do not require Ethics Committee permissions.

Informed Consent Statement: Not applicable.

Data Availability Statement: Data available on request from the authors.

Conflicts of Interest: The authors declare no conflicts of interest.

Abbreviations

DCI	D-chiro-Inositol
EI	Epi-Inositol
FET	fish embryo toxicity
hpf	hours post fertilization
MI	Myo-Inositol
PCOS	Polycystic Ovary Syndrome
SI	Scyllo-Inositol
SEM	standard error of the mean
VMR	visual motor response

References

1. Antonowski, T.; Wiśniewski, K.; Podlasz, P.; Osowski, A.; Wojtkiewicz, J. Study of the Potential Hepatoprotective Effect of Myo-Inositol and Its Influence on Zebrafish Development. *Nutrients* **2021**, *13*, 3346. [CrossRef] [PubMed]
2. Bizzarri, M.; Fuso, A.; Dinicola, S.; Cucina, A.; Bevilacqua, A. Pharmacodynamics and Pharmacokinetics of Inositol(s) in Health and Disease. *Expert Opin. Drug Metab. Toxicol.* **2016**, *12*, 1181–1196. [CrossRef] [PubMed]
3. Loewus, F.A.; Murthy, P.P.N. Myo-Inositol Metabolism in Plants. *Plant Sci.* **2000**, *150*, 1–19. [CrossRef]
4. Owczarczyk-Saczonek, A.; Lahuta, L.; Ligor, M.; Placek, W.; Górecki, R.; Buszewski, B. The Healing-Promoting Properties of Selected Cyclitols—A Review. *Nutrients* **2018**, *10*, 1891. [CrossRef] [PubMed]
5. Fontana, B.D.; Müller, T.E.; Cleal, M.; De Abreu, M.S.; Norton, W.H.J.; Demin, K.A.; Amstislavskaya, T.G.; Petersen, E.V.; Kalueff, A.V.; Parker, M.O.; et al. Using Zebrafish (Danio Rerio) Models to Understand the Critical Role of Social Interactions in Mental Health and Wellbeing. *Prog. Neurobiol.* **2022**, *208*, 101993. [CrossRef] [PubMed]
6. Daly, M.; Robinson, E. Depression and Anxiety during COVID-19. *Lancet* **2022**, *399*, 518. [CrossRef] [PubMed]
7. Santomauro, D.F.; Mantilla Herrera, A.M.; Shadid, J.; Zheng, P.; Ashbaugh, C.; Pigott, D.M.; Abbafati, C.; Adolph, C.; Amlag, J.O.; Aravkin, A.Y.; et al. Global Prevalence and Burden of Depressive and Anxiety Disorders in 204 Countries and Territories in 2020 Due to the COVID-19 Pandemic. *Lancet* **2021**, *398*, 1700–1712. [CrossRef] [PubMed]
8. Saeed, H.; Eslami, A.; Nassif, N.T.; Simpson, A.M.; Lal, S. Anxiety Linked to COVID-19: A Systematic Review Comparing Anxiety Rates in Different Populations. *Int. J. Environ. Res. Public Health* **2022**, *19*, 2189. [CrossRef]
9. Sarris, J.; Ravindran, A.; Yatham, L.N.; Marx, W.; Rucklidge, J.J.; McIntyre, R.S.; Akhondzadeh, S.; Benedetti, F.; Caneo, C.; Cramer, H.; et al. Clinician Guidelines for the Treatment of Psychiatric Disorders with Nutraceuticals and Phytoceuticals: The World Federation of Societies of Biological Psychiatry (WFSBP) and Canadian Network for Mood and Anxiety Treatments (CANMAT) Taskforce. *World J. Biol. Psychiatry* **2022**, *23*, 424–455. [CrossRef] [PubMed]
10. Wu, J.; Yan, B.; Bao, M.; Shen, J.; Zheng, P.; Wu, D.; Wang, J.; Li, Z.; Jiang, K. Early Life Exposure to Chronic Unpredictable Stress Induces Anxiety-like Behaviors and Increases the Excitability of Cerebellar Neurons in Zebrafish. *Behav. Brain Res.* **2023**, *437*, 114160. [CrossRef]
11. Ramezani, M.; Simani, L.; Karimialavijeh, E.; Rezaei, O.; Hajiesmaeili, M.; Pakdaman, H. The Role of Anxiety and Cortisol in Outcomes of Patients With COVID-19. *Basic Clin. Neurosci. J.* **2020**, *11*, 179–184. [CrossRef]
12. Ansara, E.D. Management of Treatment-Resistant Generalized Anxiety Disorder. *Ment. Health Clin.* **2020**, *10*, 326–334. [CrossRef]
13. Streba, L.; Gheonea, D.I.; Schenker, M. *Current Trends in Cancer Management*; IntechOpen: London, UK, 2019; ISBN 978-1-83880-005-5.
14. Schnörr, S.J.; Steenbergen, P.J.; Richardson, M.K.; Champagne, D.L. Measuring Thigmotaxis in Larval Zebrafish. *Behav. Brain Res.* **2012**, *228*, 367–374. [CrossRef]
15. Bühler, A.; Carl, M. Zebrafish Tools for Deciphering Habenular Network-Linked Mental Disorders. *Biomolecules* **2021**, *11*, 324. [CrossRef]
16. Rosa, J.G.S.; Lima, C.; Lopes-Ferreira, M. Zebrafish Larvae Behavior Models as a Tool for Drug Screenings and Pre-Clinical Trials: A Review. *Int. J. Mol. Sci.* **2022**, *23*, 6647. [CrossRef] [PubMed]
17. Champagne, D.L.; Hoefnagels, C.C.M.; De Kloet, R.E.; Richardson, M.K. Translating Rodent Behavioral Repertoire to Zebrafish (Danio Rerio): Relevance for Stress Research. *Behav. Brain Res.* **2010**, *214*, 332–342. [CrossRef]
18. Selderslaghs, I.W.T.; Hooyberghs, J.; De Coen, W.; Witters, H.E. Locomotor Activity in Zebrafish Embryos: A New Method to Assess Developmental Neurotoxicity. *Neurotoxicology Teratol.* **2010**, *32*, 460–471. [CrossRef]
19. Torres-Ruiz, M.; De Alba González, M.; Morales, M.; Martin-Folgar, R.; González, M.C.; Cañas-Portilla, A.I.; De La Vieja, A. Neurotoxicity and Endocrine Disruption Caused by Polystyrene Nanoparticles in Zebrafish Embryo. *Sci. Total Environ.* **2023**, *874*, 162406. [CrossRef] [PubMed]
20. Wiśniewski, K.; Antonowski, T.; Juranek, J.; Podlasz, P.; Wojtkiewicz, J. Antiepileptic Properties of Scyllo-Inositol on Pentylenetetrazol-Induced Seizures. *Int. J. Mol. Sci.* **2023**, *24*, 7598. [CrossRef]
21. Sharma, P.; Rathore, P.; Seth, S.; Kaba, L.; Bhamawat, N.; Jain, P. Insulin Mimetic Potential of Hylocereus Undatus from Extracted Myo-Inositol and Proteins. *Braz. J. Sci.* **2023**, *2*, 12–18. [CrossRef]

22. Pkhaladze, L.; Unfer, V.; Dewailly, D. Chapter 10—Use of Myo-Inositol in the Treatment of PCOS Symptoms in Adolescents. In *A Clinical Guide to Inositols*; Unfer, V., Dewailly, D., Eds.; Academic Press: Cambridge, MA, USA, 2023; pp. 151–165. ISBN 978-0-323-91673-8.
23. Rakic, D.; Jakovljevic, V.; Jovic, N.; Bicanin Ilic, M.; Dimitrijevic, A.; Vulovic, T.; Arsenijevic, P.; Sretenovic, J.; Nikolic, M.; Petrovich Fisenko, V.; et al. The Potential of SGLT-2 Inhibitors in the Treatment of Polycystic Ovary Syndrome: The Current Status and Future Perspectives. *Biomedicines* **2023**, *11*, 998. [CrossRef]
24. Unfer, V.; Carlomagno, G.; Dante, G.; Facchinetti, F. Effects of Myo-Inositol in Women with PCOS: A Systematic Review of Randomized Controlled Trials. *Gynecol. Endocrinol.* **2012**, *28*, 509–515. [CrossRef]
25. Zhao, H.; Xing, C.; Zhang, J.; He, B. Comparative Efficacy of Oral Insulin Sensitizers Metformin, Thiazolidinediones, Inositol, and Berberine in Improving Endocrine and Metabolic Profiles in Women with PCOS: A Network Meta-Analysis. *Reprod. Health* **2021**, *18*, 171. [CrossRef]
26. Siracusa, L.; Napoli, E.; Ruberto, G. Novel Chemical and Biological Insights of Inositol Derivatives in Mediterranean Plants. *Molecules* **2022**, *27*, 1525. [CrossRef]
27. Chiappelli, J.; Rowland, L.M.; Wijtenburg, S.A.; Muellerklein, F.; Tagamets, M.; McMahon, R.P.; Gaston, F.; Kochunov, P.; Hong, L.E. Evaluation of Myo-Inositol as a Potential Biomarker for Depression in Schizophrenia. *Neuropsychopharmacology* **2015**, *40*, 2157–2164. [CrossRef]
28. Shirayama, Y.; Takahashi, M.; Osone, F.; Hara, A.; Okubo, T. Myo-Inositol, Glutamate, and Glutamine in the Prefrontal Cortex, Hippocampus, and Amygdala in Major Depression. *Biol. Psychiatry Cogn. Neurosci. Neuroimaging* **2017**, *2*, 196–204. [CrossRef]
29. Coupland, N.J.; Ogilvie, C.J.; Hegadoren, K.M.; Seres, P.; Hanstock, C.C.; Allen, P.S. Decreased Prefrontal Myo-Inositol in Major Depressive Disorder. *Biol. Psychiatry* **2005**, *57*, 1526–1534. [CrossRef]
30. Cantelmi, T.; Lambiase, E.; Unfer, V.; Gambioli, R.; Unfer, V. Inositol Treatment for Psychological Symptoms in Polycystic Ovary Syndrome Women. *Eur. Rev. Med. Pharmacol. Sci.* **2021**, *25*, 2383–2389.
31. Bersudsky, Y.; Einat, H.; Stahl, Z.; Belmaker, R.H. Epi-Inositol and Inositol Depletion: Two New Treatment Approaches in Affective Disorder. *Curr. Psychiatry Rep.* **1999**, *1*, 141–147. [CrossRef] [PubMed]
32. Shaldubina, A.; Buccafusca, R.; Johanson, R.A.; Agam, G.; Belmaker, R.H.; Berry, G.T.; Bersudsky, Y. Behavioural Phenotyping of Sodium-Myo-Inositol Cotransporter Heterozygous Knockout Mice with Reduced Brain Inositol. *Genes Brain Behav.* **2007**, *6*, 253–259. [CrossRef] [PubMed]
33. Derkaczew, M.; Martyniuk, P.; Osowski, A.; Wojtkiewicz, J. Cyclitols: From Basic Understanding to Their Association with Neurodegeneration. *Nutrients* **2023**, *15*, 2029. [CrossRef]
34. Mashayekh-Amiri, S.; Delavar, M.A.; Bakouei, F.; Faramarzi, M.; Esmaeilzadeh, S. The Impact of Myo-Inositol Supplementation on Sleep Quality in Pregnant Women: A Randomized, Double-Blind, Placebo-Controlled Study. *J. Matern.-Fetal Neonatal Med.* **2022**, *35*, 3415–3423. [CrossRef] [PubMed]
35. Tariot, P.; Lyketsos, C.; Crans, G.; Cedarbaum, J.; Hernandez, C.; Abushakra, S. The Effects of ELND005 (Scyllo-Inositol) on Emergence of Neuropsychiatric Symptoms (NPS) in Mild/Moderate Alzheimer's Disease: Results from a 78-Week Phase 2 Study (P04.215). *Neurology* **2012**, *78*, P04.215. [CrossRef]
36. Einat, H.; Elkabaz-Shwortz, Z.; Cohen, H.; Kofman, O.; Belmaker, R.H. Chronic Epi-Inositol Has an Anxiolytic-like Effect in the plus-Maze Model in Rats. *Int. J. Neuropsychopharm.* **1998**, *1*, 31–34. [CrossRef] [PubMed]
37. Einat, H.; Shaldubina, A.; Belmaker, R.H. Epi-Inositol: A Potential Antidepressant. *Drug Dev. Res.* **2000**, *50*, 309–315. [CrossRef]
38. Einat, H.; Belmaker, R.H. The Effects of Inositol Treatment in Animal Models of Psychiatric Disorders. *J. Affect. Disord.* **2001**, *62*, 113–121. [CrossRef] [PubMed]
39. Alonso-Castro, A.J.; Alba-Betancourt, C.; Rocha-González, E.; Ruiz-Arredondo, A.; Zapata-Morales, J.R.; Gasca-Martínez, D.; Pérez-Gutiérrez, S. Neuropharmacological Effects of D-pinitol and Its Possible Mechanisms of Action. *J. Food Biochem.* **2019**, *43*, e13070. [CrossRef] [PubMed]
40. Fajemiroye, J.O.; Da Silva, D.M.; De Oliveira, D.R.; Costa, E.A. Treatment of Anxiety and Depression: Medicinal Plants in Retrospect. *Fundam. Clin. Pharmacol.* **2016**, *30*, 198–215. [CrossRef] [PubMed]
41. Cui, W.; Ma, A.; Huang, Z.; Liu, Z.; Yang, K.; Zhang, W. Myo-Inositol Facilitates Salinity Tolerance by Modulating Multiple Physiological Functions in the Turbot Scophthalmus Maximus. *Aquaculture* **2020**, *527*, 735451. [CrossRef]
42. De Oliveira, A.; Brigante, T.; Oliveira, D. Tail Coiling Assay in Zebrafish (Danio Rerio) Embryos: Stage of Development, Promising Positive Control Candidates, and Selection of an Appropriate Organic Solvent for Screening of Developmental Neurotoxicity (DNT). *Water* **2021**, *13*, 119. [CrossRef]
43. Endesfelder, S.; Weichelt, U.; Schiller, C.; Winter, K.; Von Haefen, C.; Bührer, C. Caffeine Protects Against Anticonvulsant-Induced Impaired Neurogenesis in the Developing Rat Brain. *Neurotox. Res.* **2018**, *34*, 173–187. [CrossRef] [PubMed]
44. Burton, C.E.; Zhou, Y.; Bai, Q.; Burton, E.A. Spectral Properties of the Zebrafish Visual Motor Response. *Neurosci. Lett.* **2017**, *646*, 62–67. [CrossRef] [PubMed]
45. Egan, R.J.; Bergner, C.L.; Hart, P.C.; Cachat, J.M.; Canavello, P.R.; Elegante, M.F.; Elkhayat, S.I.; Bartels, B.K.; Tien, A.K.; Tien, D.H.; et al. Understanding Behavioral and Physiological Phenotypes of Stress and Anxiety in Zebrafish. *Behav. Brain Res.* **2009**, *205*, 38–44. [CrossRef] [PubMed]

46. Bai, Y.; Liu, H.; Huang, B.; Wagle, M.; Guo, S. Identification of Environmental Stressors and Validation of Light Preference as a Measure of Anxiety in Larval Zebrafish. *BMC Neurosci.* **2016**, *17*, 63. [CrossRef] [PubMed]
47. Schnörr, S.J.; Steenbergen, P.J.; Richardson, M.K.; Champagne, D.L. Assessment of Thigmotaxis in Larval Zebrafish. In *Zebrafish Protocols for Neurobehavioral Research*; Kalueff, A.V., Stewart, A.M., Eds.; Neuromethods; Humana Press: Totowa, NJ, USA, 2012; pp. 37–51. ISBN 978-1-61779-597-8.

nutrients

MDPI

Article

Bioavailability of Lutein from Marigold Flowers (Free vs. Ester Forms): A Randomised Cross-Over Study to Assess Serum Response and Visual Contrast Threshold in Adults

Begoña Olmedilla-Alonso [1,*], Fernando Granado-Lorencio [2,†], Julio Castro-Feito [3], Carmen Herrero-Barbudo [2], Inmaculada Blanco-Navarro [2,†] and Rocío Estévez-Santiago [4]

1 Instituto de Ciencia y Tecnología de Alimentos y Nutrición (ICTAN-CSIC), c/José Antonio Novais, 6, 28040 Madrid, Spain

2 Hospital Universitario Puerta de Hierro-Majadahonda, c/Maestro Rodrigo, 2, 28222 Majadahonda, Spain; granadof@hotmail.es (F.G.-L.); maria.herrero@cfppuertadehierro.org (C.H.-B.); macublanco10@gmail.com (I.B.-N.)

3 Optometric Center Zires, c/Juan de Austria, 16, 28016 Madrid, Spain; jcastro@zires.es

4 Facultad de Ciencias de la Salud, Universidad Francisco de Vitoria, Ctra. Pozuelo-Majadahonda Km 1800, 28223 Pozuelo de Alarcón, Spain; rocio.estevez@ufv.es

* Correspondence: bolmedilla@ictan.csic.es; Tel.: +34-91-549-23-00

† F.G.-L. and I.B.-N. are currently retired.

Abstract: Lutein (Lut) and zeaxanthin (Zeax) are found in the blood and are deposited in the retina (macular pigment). Both are found in the diet in free form and esterified with fatty acids. A high intake and/or status is associated with a lower risk of chronic diseases, especially eye diseases. There is a large global demand for Lut in the dietary supplement market, with marigold flowers being the main source, mainly as lutein esters. As the bioavailability of Lut from free or ester forms is controversial, our aim was to assess the bioavailability of Lut (free vs. ester) and visual contrast threshold (CT). Twenty-four healthy subjects (twelve women, twelve men), aged 20–35 and 50–65 years, were enrolled in a cross-sectional study to consume 6 mg lutein/day from marigold extract (free vs. ester) for two months. Blood samples were taken at baseline and after 15, 40, and 60 days in each period. Serum Lut and Zeax were analysed using HPLC, and dietary intake was determined with a 7-day food record at the beginning of each period. CT, with and without glare, was at 0 and 60 days at three levels of visual angle. Lut + Zeax intake at baseline was 1.9 mg/day, and serum lutein was 0.36 μmol/L. Serum lutein increased 2.4-fold on day 15 (up to 0.81 and 0.90 μmol/L with free and ester lutein, respectively) and was maintained until the end of the study. Serum Zeax increased 1.7-fold. There were no differences in serum Lut responses to free or ester lutein at any time point. CT responses to lutein supplementation (free vs. ester) were not different at any time point. CT correlated with Lut under glare conditions, and better correlations were obtained at low frequencies in the whole group due to the older group. The highest correlations occurred between CT at high frequency and with glare with serum Lut and Lut + Zeax. Only in the older group were inverse correlations found at baseline at a high frequency with L + Z and with Lut/cholesterol and at a low frequency with Lut/cholesterol. In conclusion, daily supplementation with Lut for 15 days significantly increases serum Lut in normolipemic adults to levels associated with a reduced risk of age-related eye disease regardless of the chemical form of lutein supplied. Longer supplementation, up to two months, does not significantly alter the concentration achieved but may contribute to an increase in macular pigment (a long-term marker of lutein status) and thus improve the effect on visual outcomes.

Keywords: lutein intake; blood lutein; lutein free; lutein ester; contrast sensitivity; zeaxanthin

1. Introduction

Lutein and its structural isomer zeaxanthin are obtained exclusively from food. These carotenoids can be absorbed intact or undergo oxidative cleavage prior to absorption from the intestinal lumen and are transported in the blood by lipoproteins, associated with LDL and HDL lipoproteins. Lutein and zeaxanthin are preferentially deposited in the retina where, together with meso-zeaxanthin (thought to be formed from lutein in the retina), they form the macular pigment (MP). This specific intraocular deposition suggests a biological process governing their capture, deposition, and stabilisation in the macula, thought to be mediated by binding proteins [1].

Lutein and zeaxanthin are oxygenated derivatives of carotenes (xanthophylls), which are mainly provided by green plant foods, olive oil, eggs, and egg by-products [1]. In most fruits and some vegetables, lutein and zeaxanthin are found in free form and esterified with fatty acids [2–4]. Although xanthophyll esters are the main form in most fruits and flowers, there is much less information on the occurrence of carotenoid esters in foods of plant and animal origin compared to the large amount of data available on the free forms of carotenoids [5]. Moreover, in the diet, the main xanthophylls (lutein, zeaxanthin, and β-cryptoxanthin) are mainly obtained as ester forms, as described in Spanish and Brazilian populations, where they represent about 70% of the total dietary xanthophyll intake [5].

On the other hand, high dietary lutein intake and/or blood concentration is associated with a lower risk of several chronic diseases (development or progression), most notably age-related macular degeneration (ARMD) [6–8], as well as health benefits, such as improved visual and cognitive function [7,9–13]. A reduction in the risk of progression from intermediate to advanced ARMD with lutein-rich supplementation [8] and improvements in visual function have also been described [14,15]. In the eye, the MP filters (or absorbs) blue light, physically protecting the underlying photoreceptor cell layer from light damage, but also acts as an antioxidant to protect against the formation of reactive oxygen species [1,15].

Today, lutein is one of the most sought-after bioactive compounds in the global food supplement market, mainly due to its protective role in the retina, in relation to the cosmetic industry (e.g., sun light protection), and as a colour additive in food and feed [16,17]. The main commercial source of lutein for dietary supplements comes from marigold flowers (*Tagetes erecta*), which are rich in carotenoids, especially lutein, which is mainly present as lutein ester (90–99%) [4]. Although the esterification of xanthophylls seems to facilitate their bioaccessibility and the ester forms are more stable during food processing [4,18], the bioavailability of lutein from free or ester forms seems to be controversial. The final product in the blood is free lutein, as lutein esters are hydrolysed in the lumen of the small intestine before being taken up by the enterocyte [1]. However, lutein esters have also been described in the serum of normolipemic individuals who take lutein supplements and achieve serum lutein concentrations above 1.06 μmol/L (60 μg/dL) [19].

Although in general the bioavailability of xanthophyll esters seems to be equal or higher than their free forms [5,18], with regard to lutein, while some studies show that the lutein ester is better absorbed [20,21], others show the opposite, the free form being better absorbed than the ester form [22], or even no difference [23,24]. Therefore, as more studies on lutein bioavailability (free versus ester) are needed, preferably in subjects with well-defined characteristics and in studies acting as their own control, our aims were to assess lutein bioavailability, free versus ester forms, taking into account two potential confounding factors, sex and age, and to provide an approach to the assessment of visual function in the context of lutein supplementation.

2. Materials and Methods

2.1. Subjects and Study Design

A total of 24 apparently healthy subjects (12 women, 12 men) in two different age groups (20–35 and 50–65 years) were recruited according to the inclusion and exclusion criteria. Inclusion criteria were as follows: serum cholesterol $\leq$ 5.7 mmol/L, biochemical and haematological profile within normal limits, and no medication, vitamin, or dietary supplements. Exclusion criteria were as follows: habitual use of lutein supplements and cataract or any ocular disease affecting the macula. Volunteers were enrolled in a randomised cross-over study to take a lutein supplement (free or ester form) of 6 mg lutein/day for two months, followed by a two-month wash-out period (Figure 1). Participants were asked to continue their usual diet and to take lutein capsules with one of their daily meals.

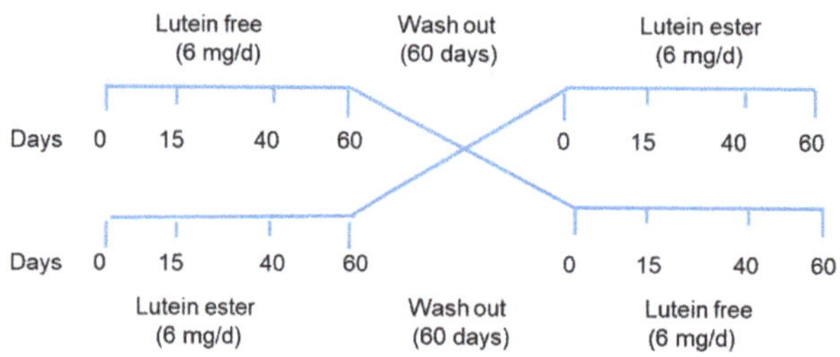

Figure 1. Study design $n = 24$, divided into two groups: lutein-free, $n = 12$ ($n = 6$, three women and three men aged 20–35 y, and $n = 6$, three women and three men aged 50–65 y) and lutein ester, $n = 12$, with the same distribution as in the lutein free group). Blood samples were taken at each time point, CT was assessed at 0 and 60 days, and food records were taken at the start of each period.

Blood samples (8–10 mL) were collected after an 8–10 h fast at baseline and at days 15, 40, and 60 of each of the two intervention periods. Subjects underwent a 7-day food record at the start of each supplementation period. The visual system was assessed using contrast threshold (CT) at 0 and 60 days. This study was approved by the Clinical Research Ethics Committee of the Hospital Universitario Puerta de Hierro of Madrid, Spain (Dated 27 April 2007, code: CEIm.), and the subjects gave their written informed consent.

2.2. Lutein Supplements

Lutein capsules (6 mg/capsule) were manufactured by Goerlich Pharma International, GmbH (Edling, Germany) using Flora GLO 10% VG TabGrade™ (Kemin, Des Moines, IA, USA) and from Xangold 10% (Cognis, Monheim am Rhein, Germany) for the capsules containing the free and ester forms of lutein, respectively. Both Flora GLO and Xangold are made from marigold flowers and contain low levels of zeaxanthin (a minimum of 0.4% in Flora GLO and approximately 4.9% in Xangold, according to the manufacturers). The overall composition of the free lutein capsules was as follows: Flora GLO 10% VG 22.22%, cellulose microcrystalline 77.04%, and magnesium stearate 2%, and that of the lutein ester capsules was: Xangold 10% 43.33%, cellulose microcrystalline 55.93%, and magnesium stearate 0.74%. The capsules were provided at each visit, with a few extra capsules each time.

2.3. Assessment of Dietary Intake of Lutein and Carotenoids

Dietary intake of carotenoids was assessed using 7-day food records at the beginning of each of the two intervention periods. Food records were kept for seven consecutive days (including one weekend) and were collected by a specialised dietician who reviewed the records in the presence of the participant, asking questions about quantities or items to facilitate correct identification of the food/drink consumed. Amount consumed was estimated in portions or household servings and in units (fruit) [25]. On the basis of this information, we calculated the dietary intake (grams/day), which was used to determine the daily intake of lutein and zeaxanthin using a database developed by our group and included in a software application for the assessment of individual dietary intake of carotenoids [26]. This database contains HPLC-generated data on the main dietary carotenoids, mainly from fruits and vegetables [27–29].

2.4. Analysis of Serum Concentrations of Lutein and Zeaxanthin

Serum concentrations of lutein and zeaxanthin were determined using HPLC using a system consisting of a model 6000 pump, a Rheodyne injector and a 2998 photodiode array detector (Waters, Milford, MA, USA), and a Spheri-5 ODS 5 μm column (220 mm × 4.6 mm (Brownlee Labs, Applied Biosystem, Santa Clara, CA, USA) with a guard column (Aquapore ODS type RP-18, PerkinElmer Inc., Drachten, The Netherlands). The mobile phases were acetonitrile–methanol (85:15; v/v), and acetonitrile–dichloromethane–methanol (70:20:10; $v/v/v$) in a linear gradient from 5 min to 20 min. Both mobile phases were stabilised with ammonium acetate (0.025 mol L^{-1}) added to the methanol. The flow rate was 1.8 mL min^{-1}, and detection was at a wavelength of 450 nm. All chromatograms were processed using Empower 2 software (Waters, Milford, MA, USA).

Blood samples were collected in plain blood collection tubes without anticoagulant. The serum was separated using centrifugation (630× g, 10 min). Baseline and supplemented serum samples from each subject were stored at 70 °C, processed, and analysed simultaneously to reduce analytical variability. Subjects were analysed in random order within seven months of collection, and one in six serum samples were analysed in duplicate. Carotenoid extraction was performed on fasting serum samples using a slight modification of a previously published method [30]. Briefly, 200 μL of serum was added to 200 μL of ethanol, vortexed, and extracted twice with 400 μL of hexane: dichloromethane (5:1) stabilised with 0.1 g/L butylated hydroxyltoluene. The organic phases were pooled, evaporated to dryness under a nitrogen atmosphere, reconstituted with 200 μL of a solution of tetrahydrofuran: ethanol (1:2), and injected (5 μL) into the HPLC system. Methanol, ethanol, acetonitrile, dichloromethane, ammonium acetate, butylated hydroxytoluene (BHT), and tetrahydrofuran were supplied by Panreac (Barcelona, Spain). Lutein (xanthophyll from marigold) was purchased from Sigma Chemical Co. (St. Louis, MO, USA), and zeaxanthin was purchased from Fluka Analytica (Sigma Aldrich, St. Louis, MI, USA).

Blood total cholesterol was analysed with colourimetric enzymatic assay (Cobas Integra 400 plus, Roche, Basel, Switzerland).

2.5. Visual Contrast Threshold

Visual function was assessed by measuring contrast threshold (CT) with and without glare (Contrast Glaretester, CGT-1000, Takagi Sciko Co. Ltd., Nagano, Japan) at three times during the study period (0, 40, 60 days). The CGT-1000 determined CT with and without glare using an automated strategy set for 6 sizes of annular stimuli with diameters ranging from 6.3° to 0.7° of visual angle. The 12 levels of CT range from 0.01 to 0.45. The luminance of the background on which the stimuli were presented was 10 cd/m^2. Stimulus presentation lasted 0.2 s with 0.2 s intervals; test luminance had a glare of 40,000 cd/m^2.

The lower the CT, the higher the contrast sensitivity level at which a subject was able to detect each spatial frequency. The reciprocal of the CT is called contrast sensitivity. Each subject was tested monocularly for CT, once with each eye and with spectacle correction if necessary. The test results were automatically printed on a single graph showing the sensitivity functions. CT results were reported as the mean of three levels of frequency data—low, medium, high—corresponding to the mean of 6.3° and 4° of visual angle (lower frequencies), 2.5° and 1.6° (medium frequencies), and 1.0° and 0.7° (high frequencies).

2.6. Statistical Analysis

Data are expressed as the mean and standard error, 95% confidence interval. The normal distribution of the data was assessed (Kolmogorov–Smirnov test). Serum lutein did not follow a normal distribution at any time point in the lutein-free group and at baseline in the ester group. CT and lutein data from both groups showed a normal distribution at baseline but not at 60 days, so Pearson and Spearman correlations were used. Non-parametric tests (Friedman) were used to compare the concentrations of the variables analysed in the groups (four groups according to sex and age), the Willcoxon test to compare the concentrations at baseline, and the Kruskall–Wallis test to compare the responses to supplementation. Although there were differences for sex and age in each of the four groups, no interactions were observed for these variables in the lutein chemical formula groups. A linear mixed model with repeated measures for time and lutein chemical formula was used with lutein and lutein/cholesterol as dependent variables and age, sex, time, and chemical formula as fixed factors. Pairwise interactions of all factors were analysed.

All reported p-values are based on a two-tailed test and compared at a 5% significance level. IBM SPSS Statistics, v.27 software was used for all statistical calculations.

3. Results

The subjects (n = 24, divided into two age groups: n = 12 are 20–35 years old (mean and SD: 25 ± 3 y) and n= 12 are 50–65 years old (mean and SD: 53 ± 4 y)) supplemented their usual daily diet with 6 mg of lutein (free and ester), which was three times their mean dietary intake of lutein + zeaxanthin at a baseline of 1.9 mg/day (±0.23 SE) (median = 1.4 mg/day). Compliance was assessed by counting the number of capsules returned at each visit, and it was greater than 85%. As the lutein capsules supplied also contained zeaxanthin, albeit at a very low concentration compared to lutein, serum zeaxanthin was also analysed. Table 1 shows the serum concentrations of lutein, lutein/cholesterol, and zeaxanthin at baseline and at each time point analysed during the study (n = 48). There were no differences in the concentrations of lutein, zeaxanthin, and cholesterol (mean and SD: 4.69 ± 0.90 and 4.71 ± 0.89 mmol/L) at the beginning of each of the supplementation periods.

Table 1. Serum lutein, zeaxanthin (µmol/L), and lutein/cholesterol concentrations in participants, expressed as mean (SE) [95%CI] throughout the study.

	Free Lutein Group (n = 24)			**Lutein Ester Group (n = 24)**			**Total Group (n = 48)**		
	Lutein	**Lutein/Cholesterol**	**Zeaxanthin**	**Lutein**	**Lutein/Cholesterol**	**Zeaxanthin**	**Lutein**	**Zeaxanthin**	**Lutein + Zeaxanthin**
Basal	0.35 ^{a} (0.03) [0.29, 0.42]	0.11 ^{a} (0.01) (0.09, 0.13]	0.09 ^{a} (0.01) [0.07, 0.11]	0.38 ^{a} (0.03) [0.32, 0.44]	0.12 ^{a} (0.47) [0.10, 0.14]	0.10 ^{a} (0.01) [0.08, 0.12]	0.37 ^{a} (0.03) [0.31, 0.42]	0.10 ^{a} (0.01) [0.08, 0.11]	0.51 ^{a} (0.04) [0.43, 0.59]
15 d	0.81 ^{b} (0.04) [0.72, 0.89]	0.26 ^{b} (0.02) [0.22, 0.29]	0.14 ^{b} (0.01) [0.12, 0.16]	0.90 ^{b} (0.06) [0.79, 1.01]	0.29 ^{b} (0.02) [0.25, 0.33]	0.14 ^{b} (0.01) [0.13, 0.16]	0.85 ^{b} (0.04) [0.78, 0.93]	0.14 ^{b} (0.01) [0.13, 0.15]	1.02 ^{b} (0.05) [0.92, 1.12]
40 d	0.87 ^{b} (0.06) [0.74, 1.00]	0.29 ^{b} (0.02) [0.24, 0.34]	0.16 ^{bc} (0.01) [0.13, 0.18]	0.94 ^{b} (0.05) [0.84, 1.04]	0.31 ^{b} (0.02) [0.27, 0.36]	0.16 ^{bc} (0.01) [0.14, 0.18]	0.91 ^{b} (0.04) [0.83, 0.99]	0.16 ^{c} (0.01) [0.14, 0.17]	1.10 ^{c} (0.05) [1.03, 1.20]
60 d	0.84 ^{b} (0.07) [0.69, 0.99]	0.27 ^{b} (0.03) [0.22, 0.33]	0.16 ^{b} (0.01) [0.13, 0.18]	0.96 ^{b} (0.06) [0.84, 1.07]	0.31 ^{b} (0.02) [0.27, 0.04]	0.17 ^{b} (0.01) [0.15, 0.19]	0.90 ^{b} (0.05) [0.80, 1.00]	0.16 ^{c} (0.01) [0.15, 0.18]	1.08 ^{c} (0.06) [0.96, 1.20]

^{a,b,c} In the columns, different superscripts indicate statistical significance (p < 0.05).

The serum lutein concentration increased on day 15, reaching levels of 0.81 and 0.90 µmol/L for free and ester lutein, respectively. These increases, which averaged 2.4 times, were maintained throughout the intervention study (days 40 and 60) in each group. The serum zeaxanthin concentration also increased from 0.10 to 0.16 µmol/L, an average 1.7-fold increase (Table 1). Zeaxanthin increased on day 15 (p < 0.001) and continued to increase on day 40 (p < 0.034), and it was maintained until the end of the study (60 d). At each time point, there were no differences between the responses to lutein and zeaxanthin supplementation with the two chemical formulae.

In the whole sample, serum lutein concentrations differed between sexes (0.44 ± 0.24 and 0.31 ± 0.12 µmol/L, men and women, respectively, p = 0.026) and age groups (0.27 ± 0.07 and 0.48 ± 0.23 µmol/L, younger and older, respectively, p = 0.011) at the beginning of the two supplementation periods and on day 15 but not after 40 and 60 days of supplementation. Age and sex had a weak effect on the serum lutein response to the supplementation (free or ester lutein), with higher responses in men than in women and in the older group (50–65 years), although this did not reach statistical significance. Figure 2 shows the time course of serum lutein concentration using the lutein chemical formula for groups according to sex and age.

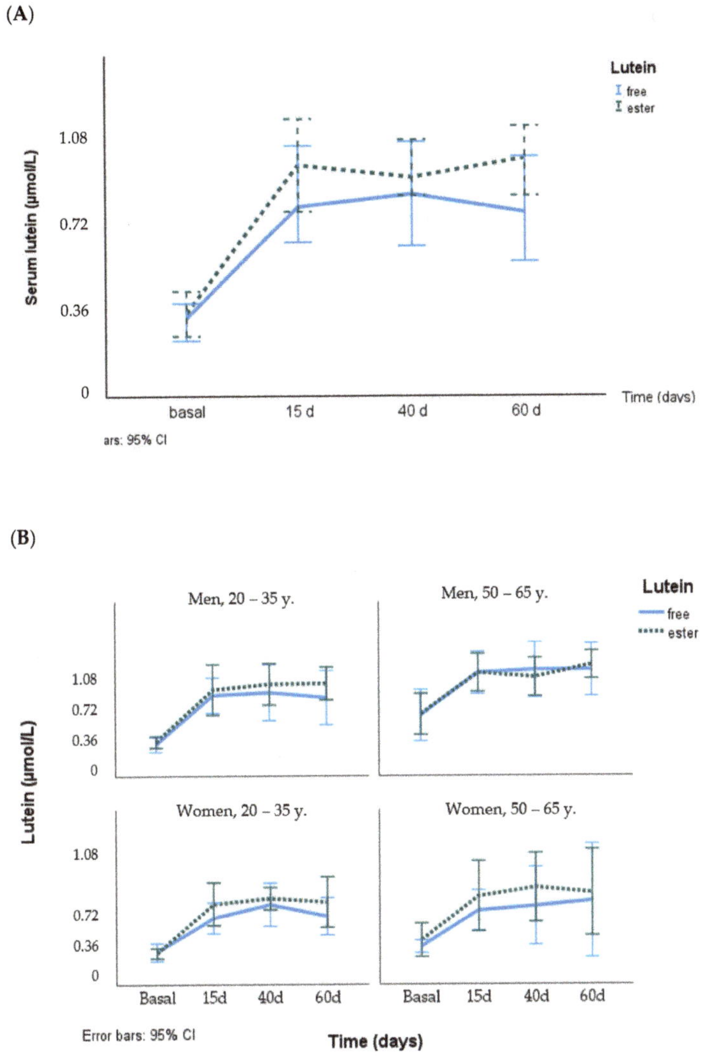

Figure 2. Response of serum lutein concentration (µmol/L) to supplementation with free lutein and lutein esters in the total group (**A**) and grouped by age and sex (**B**).

The CT data (mean ± SD, [median]), with and without glare, for the six different degrees of visual angle, grouped into three levels according to the frequencies low (6.3° and 4°), medium (2.5° and 1.6°), and high (1.0° and 0.7°), are shown in Table 2. In the total group and also grouped by age, CT showed no significant differences at baseline in any of the periods with free lutein and lutein ester, and no differences were found in the responses to lutein supplementation with the two chemical formulae at any time. CT showed statistical differences according to light level, being higher in the presence of medium- and high-frequency glare at both baseline and 60 days and also in the presence of low-frequency glare at 60 days (Table 2 and Figure 3). Although CT differed by age at medium and high frequencies, there were no differences after adjustment for age (20–25 vs. 50–65 years). CT showed no differences after 60 days of lutein supplementation except in the glare condition for the low frequencies (p = 0.008).

Table 2. Contrast thresholds (mean ± SD [median]) at different degrees of visual angle, with and without glare (n = 96 eyes), at baseline and at the end of the study.

	Contrast Threshold		
Visual Angle of the Estimulus (°)	without Glare	with Glare	
	Baseline	Baseline	p-value
6.3° and 4.0° (mean)	0.013 ± 0.013 [0.010]	0.013 ± 0.008 [0.010] [a]	0.145
2.5° and 1.6 (mean)	0.026 ± 0.038 [0.029]	0.037 ± 0.056 [0.017]	<0.001
1.0° and 0.7° (mean)	0.108 ± 0.108 [0.050]	0.118 ± 0.109 [0.060]	0.009
	60 days	60 days	
6.3° and 4.0° (mean)	0.016 ± 0.038 [0.010]	0.014 ± 0.011 [0.010] [b]	0.050
2.5° and 1.6 (mean)	0.027 ± 0.044 [0.016]	0.033 ± 0.046 [0.020]	<0.001
1.0° and 0.7° (mean)	0.105 ± 0.010 [0.055]	0.112 ± 0.105 [0.070]	0.040

Within columns, different superscripts indicate significant difference between baseline and 60 days ([ab] $p = 0.008$).

(**A**)

(**B**)

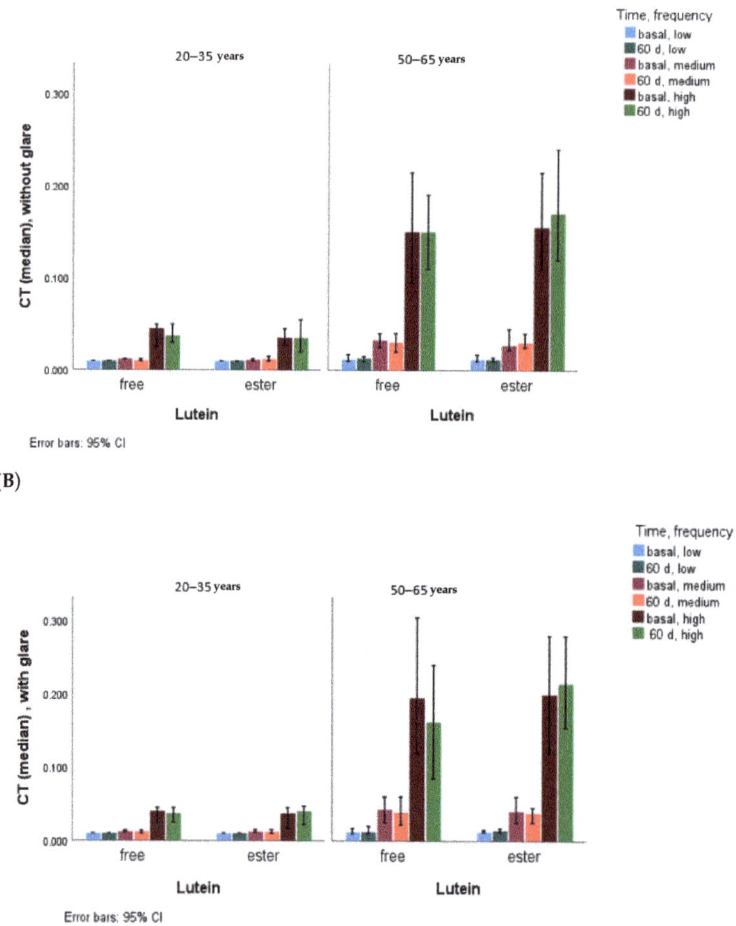

Figure 3. CT values without glare (**A**) and with glare (**B**) according to age and chemical form of lutein. Statistical differences ($p < 0.001$) at all frequency levels, with and without glare, at baseline and at 60 days between younger and older subjects.

After 60 d of lutein supplementation, age-adjusted CT was different in the glare condition for the low frequencies ($p = 0.008$).

Dietary lutein plus zeaxanthin intake showed no significant correlation with serum lutein, lutein + zeaxanthin, and lutein/cholesterol concentrations or with the CT at baseline. For serum concentrations of lutein plus zeaxanthin, lutein, and lutein/cholesterol, their correlations with CT, without and with glare, at baseline and after 60 days of supplementation in the total group (both lutein groups, free and ester) are shown in Table 3. In the total group, statistically significant correlations were found at baseline between serum lutein or lutein plus zeaxanthin concentrations and CT at each of the three frequency levels, with and without glare, but no significant correlations were found between CT and lutein/cholesterol concentrations. After 60 days of lutein supplementation, in the total group, with and without glare, all correlations of CT with lutein plus zeaxanthin and also with lutein were maintained except those under glare at low frequency and without glare at high frequency. The highest correlations were observed with glare and at the highest frequencies (lower visual angle of the estimulus) with serum lutein (0.322, $p = 0.001$) and lutein plus zeaxanthin (0.340, $p < 0.001$).

Table 3. Correlations (Spearman's rho and [p-value]) of contrast thresholds (mean at each of the three frequency levels, with and without glare) with serum lutein, lutein plus zeaxanthin, and lutein/cholesterol in the total group. Statistically significant results are shown in italics.

Visual Angle of the Estimulus (°)	Lutein	Lutein + Zeaxanthin	Lutein/Cholesterol	Lutein	Lutein + Zeaxanthin	Lutein/Cholesterol
	Baseline—without glare			Baseline—with glare		
6.3° and 4.0°	*0.345 [0.001]*	*0.329 [0.002]*	0.059 [0.591]	*0.263 [0.010]*	*0.220 [0.031]*	−0.001 [0.989]
2.5° and 1.6	*0.376 [<0.001]*	*0.293 [0.007]*	0.063 [0.569]	*0.428 [<0.001]*	*0.376 [<0.001]*	0.120 [0.245]
1.0° and 0.7°	*0.364 [<0.001]*	*0.264 [0.015]*	0.086 [0.434]	*0.359 [<0.001]*	*0.325 [0.001]*	0.110 [0.284]
	60 days—without glare			60 days–with glare		
6.3° and 4.0°	*0.239 [0.026]*	*0.231 [0.032]*	0.019 [0.862]	0.188 [0.067]	*0.210 [0.040]*	−0.031 [0.762]
2.5° and 1.6	*0.262 [0.015]*	*0.276 [0.010]*	−0.039 [0.720]	*0.213 [0.037]*	*0.254 [0.013]*	−0.061 [0.554]
1.0° and 0.7°	0.170 [0.117]	*0.232 [0.032]*	−0.095 [0.386]	*0.322 [0.001]*	*0.340 [<0.001]*	0.062 [0.547]

Comparing the age groups, at baseline in only the older group and in the glare condition, inverse correlations were found at high frequencies with lutein plus zeaxanthin (−0.343, $p = 0.017$) and with lutein/cholesterol (−0.390, $p = 0.006$) and at low frequencies with lutein/cholesterol (−0.326, $p = 0.024$). Instead, after 60 days of lutein supplementation, correlations were found only in the younger group (20–30 years), while no significant correlations were found in the older group. In the younger group, significant correlations were obtained without glare at low and medium frequencies with serum lutein (0.308, $p = 0.035$; 0.315, $p = 0.021$, respectively) and with glare at high frequency with serum lutein (0.367, $p = 0.010$) and with serum lutein/cholesterol (0.423, $p = 0.003$). Figure S1 shows the correlations of serum lutein with CT, with and without glare, according to the age of the subjects at baseline and at the end of the lutein supplementation period, and an improvement can be observed at high and medium frequencies, mainly with glare.

4. Discussion

4.1. Serum Lutein and Zeaxanthin and Effect of Lutein Supplementation (Free vs. Ester)

This study of the bioavailability of lutein from dietary supplements focused on the forms of lutein (free vs. ester) found in commercially available supplements. Both forms are derived from marigold flowers and therefore also contain its structural isomer, zeaxanthin (ester forms), although at much lower concentrations [31]. The concentration in the supplements was set at 6 mg lutein/day because this amount has been associated with a reduced risk of several chronic diseases [6,32–34], although there is evidence that lower levels of lutein and zeaxanthin may be sufficient to protect against the progression of age-related macular disease [35]. The amount of 6 mg lutein can be obtained from the diet [33] by choosing lutein-rich foods (e.g., about 100 g of spinach), but it can be difficult to maintain on a regular basis, and its bioavailability from fruits and vegetables is low and generally lower than that from supplements [17].

The 6 mg lutein/day provided is much higher than the mean dietary intake of lutein + zeaxanthin consumed by the volunteers in this study (1.9 mg/day), which is similar to data from different adult populations/groups from Europe, the USA, and other countries worldwide, with ranges from 0.1 to 4.8 mg/day (mean of 2.2 mg/day), with 1.7 mg/day being the mean of the data from Spain [36]. These dietary intakes of lutein plus zeaxanthin corresponded to a blood concentration range of 0.20–0.56 μmol/L (mean 0.33 μmol/L) [36], similar to the range (0.43–0.58 μmol/L) but lower than the mean (0.51 μmol/L) found at the baseline in our study. Dietary and blood lutein concentrations are higher than those of zeaxanthin, being about four times higher in the blood in this study, which is consistent with the 4 to 5 ratio described in people from different countries [36]. In the Spanish diet, lutein is about 11 to 13 times higher than zeaxanthin [37,38].

The daily capsules contained a marigold flower extract rich in lutein (free or ester forms) and with a much lower concentration of zeaxanthin (companies reported from 0.4 to around 4.9%). Their consumption led to an increase in both xanthophylls in the serum, reaching a plateau after 15 days for lutein and 40 days for zeaxanthin. A similar plateau has been reported to be dose-dependent in some studies [39] but not in others [40,41]. Our data on serum lutein responses (6 mg/d resulted in an increase of 2.4) are comparable to those found in similar studies in healthy subjects, e.g., supplementation with 15 mg/d (extracted from marigold) increased serum lutein 5-fold and zeaxanthin approximately 2-fold [42] and supplementation with 5 mg lutein (ester from Tagetes erecta) increased serum lutein 2.6-fold [39]; however, this differs from those reported by Machida et al. [41], who found that a daily intake of 12 mg lutein (free form) for 16 weeks provoked a maximum two-fold increase in serum.

Lutein supplementation increased serum lutein concentrations to around 0.88 μmol/L, which is above the 0.60 μmol/L concentration associated with a lower risk of chronic eye disease in epidemiological studies [32,33]. In healthy individuals, this serum lutein response is influenced by inter-individual variability and the large variability in bioavailability due to the carotenoid source and host factors [36]. Among the potential confounding factors, blood lipid concentrations are not always considered inclusion or exclusion criteria, although lutein and zeaxanthin are transported by lipids and could therefore condition the response [37]. Other confounding factors, such as sex and/or age, have been described in observational studies [36,37], and the literature generally shows results adjusted for sex and/or age. However, some studies also reported no influence of these factors on the blood lutein response [20,22,39,42], which is consistent with our data, as the serum lutein concentration differed between sexes and age groups at baseline and in the serum lutein response until the plateau was reached, but their effect did not reach statistical significance, even when the increase was expressed as lutein/cholesterol.

Lutein and zeaxanthin are ingested in the diet in free or in ester form. While they are found in free form in green leaves (vegetables), they are esterified with many fatty acids in some leaves and tubers and especially in many ripe fruits [1]. In general, the bioavailability of the ester forms seems to be higher than that of the free forms due to a better solubilisation and extraction during digestion [4,18]. Therefore, a higher bioavailability could be expected from the intake of lutein supplements as the ester form compared to the free form. However, information on the carotenoid ester content in foods and food supplements is scarce [1,5]. On the other hand, commercially available lutein-rich food supplements contain lutein in both forms, free or ester, but few studies have evaluated their bioavailability in healthy subjects, and moreover, they have shown discrepant results. A higher response to lutein ester vs. free supplementation was described in some of the studies [20,21,24], but in only one of them, the difference was statistically significant [21]. In our study, a higher serum lutein concentration was also obtained in response to lutein ester supplementation but not significantly different from that obtained with free lutein. The opposite, a higher response to the free lutein supplementation, has only been reported in one study [22]. These studies differed mainly in the concentration of lutein administered (10, 12.5, and 20 mg/day), design (randomised, parallel) [20], study duration (single dose, 1 to 6 months), and sample size (most of them around 20 and the one, which showed a higher response to free lutein, with 72 subjects). In the present study, with a crossover design to minimise inter-individual variability in responses and with a lower amount of lutein supplied (closer to that present in food intake), the serum lutein response was proportional to those from the above-mentioned studies. In this study, serum lutein responses are independent of age and sex and almost independent of serum lipids, as described in other studies [20,22].

With regard to zeaxanthin, which was also provided in this study at much lower concentrations than lutein, a higher blood response to esterified than to free zeaxanthin has been described [43].

On the other hand, the dissolution of the lutein formulation has an important influence on its bioavailability [21]. For example, Bowen et al. [20] used a single dose of lutein (free and ester forms with different formulations: crystalline vs. powder) and concluded that the lutein diester formulation was more bioavailable than the free formulation but that the difference in formulation could have led to different dissolution in emulsion lipid droplets and ultimately different bioavailability.

4.2. CT and the Effect of Lutein Supplementation

Lutein and zeaxanthin are transported in the blood by lipoproteins and deposited in the retina, where they form, together with mesozeaxanthin, the macular pigment (MP), which acts as a blue light filter and can be considered a marker of long-term dietary exposure. The MP increases with lutein and zeaxanthin supplementation and is associated with improvements in contrast sensitivity and visual performance [15,40,44–48] in a dose-dependent manner (at each of the spatial frequencies of 3, 6, and 12 cycles/degrees) in healthy subjects [46] and in subjects with impaired visual function [9]. In this study and mainly under glare conditions, in the whole sample and grouped by age, as the serum lutein concentration increases, the CT value decreases, especially in the medium and high frequencies (Figure 3), and therefore, the contrast sensitivity increases.

CT was influenced by lutein supplementation, mainly under glare conditions and at low frequencies, and as in previous studies, this relationship appears to be age-dependent [47], found only in the older group, where an increase in the CT (lower contrast sensitivity) was obtained after two months of lutein supplementation. However, in the study by Yao et al. [49], a slight trend towards an improvement in the glare sensitivity (mainly at medium frequencies) was observed from the first to the third month of supplementation (20 mg/day), and only after six months of lutein supplementation was a significant change observed in healthy subjects. Also, in young subjects without chronic eye disease, Ma et al. [50] reported a trend toward higher thresholds at low and medium frequencies after three months of supplementation with 6 mg lutein/day but without improvement in glare conditions. Another study in healthy subjects (20–69 years) supplemented with 12 mg lutein/day for four months showed an improvement in contrast sensitivity (low frequency) and glare sensitivity (high and medium frequency) [41]. In subjects with age-related macular degeneration, lutein supplementation at different doses (6, 10, and 20 mg/d) with visual function assessments at six, nine, and twelve months showed no significant changes until the twelve months [15,51].

Under glare conditions, CT values were higher (worse) than without glare at medium and high frequencies, as previously described in healthy subjects of similar ages [47]. In our study, the effect of lutein supplementation is mainly observed at the medium and high frequencies in the older group, but it did not reach statistical significance, probably because of the large variability of CT values [41,47,49–51] (higher in the older group, as previously described [47]) and because of the duration of the supplementation period (two months), which may have been too short to obtain variations in healthy subjects [41,49,50], although a ceiling in serum lutein concentrations was reached before the first month of supplementation. The sample size may also have been relatively small, although a similar sample size was used in another study to obtain variations in contrast sensitivity [15].

On the other hand, although there were significant correlations between serum lutein markers (lutein, lutein plus zeaxanthin) and CT, with and without glare, at the three frequency levels in the total group, both at baseline and after supplementation, these correlations differed when comparing age groups. At baseline, inverse correlations between CT and lutein plus zeaxanthin or lutein/cholesterol (high frequency) and between CT and lutein/cholesterol (low frequency) were found only in the older group and with glare, implying an improvement in contrast sensitivity associated with higher serum lutein and zeaxanthin concentrations. This is in line with previous data in subjects with similar characteristics, where inverse correlations were also obtained at the three frequency levels, but without glare conditions [47]. This desirable association between CT and serum lutein plus zeaxanthin at baseline in the older group did not occur after the lutein supplementation period, as no association was found in the older group. In contrast, high-contrast sensitivity has been described in older women taking lutein plus zeaxanthin supplements (at various doses from 1 mg/d) for over fifteen years in eyes without pathology [48]. However, in the younger group, where no associations were found at baseline, there were several significant associations after lutein supplementation, with and without glare, with lutein and lutein/cholesterol at low, medium, and high frequencies. We do not have an explanation for these different responses to lutein supplementation in younger and older subjects, but circulating lipids (although cholesterolaemia was within the normal range in this study), LDL, different proportions of circulating HDL and LDL cholesterol, possible oxidative modifications of LDL and HDL cholesterol [52], and variations in the lipoprotein receptors that may be selective for different ratios of LDL and HDL [53] and may have an effect on retinal pigment epithelial cells cannot be excluded.

5. Conclusions

We can conclude that a daily intake of 6 mg of lutein, from marigold lutein extracts (free and ester forms), for 15 days significantly increased serum lutein concentrations in normolipemic adults to levels associated with a lower risk of age-related eye disease, regardless of the chemical form of lutein supplied. Zeaxanthin, which is also present in marigold extract but at much lower concentrations, also produced a significant increase in its serum concentration. Longer supplementation, up to two months, does not significantly modify the serum levels achieved, as serum lutein reaches a plateau concentration but may contribute to an increase in MPOD (long-term marker of lutein status) and thus improve the effect on the visual outcomes, especially in older subjects, in whom inverse strong associations between serum lutein and CT, mainly in glare conditions, were observed in this study.

Supplementary Materials: The following supporting information can be downloaded at: https://www.mdpi.com/article/10.3390/nu16101415/s1, Figure S1: Correlations between serum lutein concentration (μg/dL, X axis) and CT (Y axis) in the total group (A), in the group aged 20–35 y (B) and 50–65 y (C). Line colours represent frequencies: high (red), medium (green) and low (blue).

Author Contributions: Conceptualization, B.O.-A.; methodology, B.O.-A. and F.G.-L.; software, I.B.-N.; validation, B.O.-A. and F.G.-L.; formal analysis, J.C.-F., C.H.-B. and R.E.-S.; investigation, B.O.-A., F.G.-L. and R.E.-S.; data curation, B.O.-A. and I.B.-N.; writing—original draft preparation, B.O.-A.; writing—review and editing, B.O.-A.; project administration, B.O.-A.; funding acquisition, B.O.-A. All authors have read and agreed to the published version of the manuscript.

Funding: This research was funded by the Spanish National Research Council (CSIC), grant P. Intramural Especial 2006701208.

Institutional Review Board Statement: The study was conducted in accordance with the Declaration of Helsinki and approved by the Ethics Committee of the Hospital Universitario Puerta de Hierro of Madrid (dated 27 April 2007).

Informed Consent Statement: Informed consent was obtained from all subjects involved in the study.

Data Availability Statement: The original contributions presented in the study are included in the article/Supplementary Materials, further inquiries can be directed to the corresponding author.

Acknowledgments: The authors thank Belén Pérez-Sacristán of the Hospital Universitario Puerta de Hierro (Madrid, Spain) for her help with the blood analysis and Laura Barrios and Belén Garzón-García of the Secretaría General Adjunta de Informática of the CSIC for the statistical evaluation of the study. B. Olmedilla-Alonso is a member of the Spanish Carotenoid Network (CaRed), grant RED2022-134577-T, funded by MCIN/AEI/10.13039/501100011033.

Conflicts of Interest: The authors declare no conflicts of interest.

References

1. Rodríguez-Concepción, M.; Ávalos, F.J.; Bonet, M.L.; Boronat, A.; Gómez-Gómez, L.; Hornero-Méndez, D.; Limón, C.; Meléndez-Martinez, A.J.; Olmedilla-Alonso, B.; Palou, A.; et al. A global perspective on carotenoids: Metabolism, biotechnology, and benefits for nutrition and health. *Prog. Lipid Res.* **2018**, *70*, 62–93. [CrossRef] [PubMed]
2. Britton, G.; Khachik, F. Carotenoids in food. In *Carotenoids—Nutrition and Health*; Britton, G., Liaaen-Jensen, S., Pfander, H., Eds.; Birkhauser: Basel, Switzerland, 2009; Volume 5, pp. 45–66.
3. Mercadante, A.Z.; Rodrigues, D.B.; Petry, F.C.; Mariutti, L.R.B. Carotenoid esters in foods—A review and practical directions on analysis and occurrence. *Food Res. Int.* **2017**, *99*, 830–850. [CrossRef] [PubMed]
4. Hornero-Méndez, D. Ocurrence of carotenoid esters in foods. In *Carotenoid Esters in Foods: Physical, Chemical and Biological Properties*; Mercadante, A.Z., Ed.; The Royal Society of Chemistry: London, UK, 2019; pp. 182–284.
5. Olmedilla-Alonso, B.; Estévez-Santiago, R. Dietary intake of carotenoids: Nutritional status assessment and the importance of considering free and esters forms in foods. In *Carotenoid Esters in Foods: Physical, Chemical and Biological Properties*; Mercadante, A.Z., Ed.; The Royal Society of Chemistry: London, UK, 2019; pp. 375–389.
6. Seddon, J.M.; Ajani, U.A.; Sperduto, R.D.; Hiller, R.; Blair, N.; Burton, T.C.; Farber, M.D.; Gragoudas, E.S.; Haller, J.; Miller, D.T.; et al. Dietary carotenoids, vitamins A, C, and E, and advanced age-related macular degeneration. Eye disease case-control study group. *JAMA* **1994**, *272*, 1413–1420. [CrossRef] [PubMed]
7. Bernstein, P.S.; Li, B.; Vachali, P.P.; Gorusupudi, A.; Shyam, R.; Henriksen, B.S.; Nolan, J.M. Lutein, zeaxanthin, and meso-zeaxanthin: The basic and clinical science underlying carotenoid-based nutritional interventions against ocular disease. *Prog. Retin Eye Res.* **2016**, *50*, 34–66. [CrossRef] [PubMed]
8. Chew, E.Y.; Clemons, T.E.; Agrón, E.; Domalpally, A.; Keenan, T.D.L.; Vitale, S.; Weber, C.; Smith, D.C.; Christen, W.; AREDS2 Research Group. Long-term Outcomes of Adding Lutein/Zeaxanthin and ω-3 Fatty Acids to the AREDS Supplements on Age-Related Macular Degeneration Progression: AREDS2 Report 28. *JAMA Ophthalmol.* **2022**, *140*, 692–698. [CrossRef] [PubMed]
9. Olmedilla, B.; Granado, F.; Blanco, I.; Vaquero, M.; Cajigal, C. Lutein in patients with cataracts and age-related macular degeneration: A long-term supplementation study. *J. Sci. Food Agric.* **2001**, *81*, 904–909. [CrossRef]
10. Hammond, B.R. Dietary carotenoids and the nervous system. *Foods* **2015**, *4*, 698–701. [CrossRef] [PubMed]
11. Mares, J. Lutein and zeaxanthin isomers in eye health and disease. *Annu. Rev. Nutr.* **2016**, *36*, 571–602. [CrossRef] [PubMed]
12. Lindbergh, C.A.; Renzi-Hammond, L.M.; Hammond, B.R.; Terry, D.P.; Mewborn, C.M.; Puente, A.N.; Miller, L.S. Lutein and Zeaxanthin influence brain function in older adults: A randomized controlled trial. *J. Int. Neuropsychol. Soc.* **2018**, *24*, 77–90. [CrossRef]
13. Hammond, B.R.; Renzi-Hammond, L. The influence of the macular carotenoids on women's eye and brain health. *Nutr. Neurosci.* **2023**, *26*, 720–726. [CrossRef]
14. Richer, S.; Stiles, W.; Statkute, L.; Pulido, J.; Frankowski, J.; Rudy, D.; Pei, K.; Tsipursky, M.; Nyland, N.R. Double-masked, placebo-controlled, randomized trial of lutein and antioxidant supplementation in the intervention of atrophic age-related macular degeneration: The Veterans LAST study (Lutein Antioxidant Supplementation Trial). *Optometry* **2004**, *75*, 216–229. [CrossRef]
15. Bartlett, H.E.; Eperjesi, F. Effect of lutein and antioxidant dietary supplementation on contrast sensitivity in age-related macular disease: A randomized controlled trial. *Eur. J. Clin. Nutr.* **2007**, *61*, 1121–1127. [CrossRef] [PubMed]
16. Allied Market Research©. Lutein and Zeaxanthin Market by Type (Lutein, Zeaxanthin), by Application (Supplements, Cosmetics, Pharmaceuticals, Others): Global Opportunity Analysis and Industry Forecast, 2021–2031. November 2022. Available online: https://www.alliedmarketresearch.com/lutein-and-zeaxanthin-market (accessed on 7 January 2024).
17. Ochoa Becerra, M.; Mojica Contreras, L.; Hsieh Lo, M.; Mateos Díaz, J.; Castillo Herrera, G. Lutein as a functional food ingredient: Stability and bioavailability. *J. Funct. Foods* **2020**, *66*, 103771. [CrossRef]
18. Bunea, A.; Socaciu, C.; Pintea, A. Xanthophyll esters in fruits and vegetables. *Not. Bot. Horti Agrobot.* **2014**, *42*, 310–324. [CrossRef]
19. Granado, F.; Olmedilla, B.; Gil-Martínez, E.; Blanco, I. Lutein ester in serum after lutein supplementation in human subjects. *Brit. J. Nutr.* **1998**, *80*, 445–449. [CrossRef]

20. Bowen, P.E.; Herbst-Espinosa, S.M.; Hussain, E.A.; Stacewicz-Sapuntzakis, M. Esterification does not impair lutein bioavailability in humans. *J. Nutr.* **2002**, *132*, 3668–3673. [CrossRef]
21. Landrum, J.; Bone, R.; Mendez, V.; Valenciaga, A.; Babino, D. Comparison of dietary supplementation with lutein diacetate and lutein: A pilot study of the effects on serum and macular pigment. *Acta Biochim. Pol.* **2012**, *59*, 167–169. [CrossRef] [PubMed]
22. Norkus, E.P.; Norkus, K.L.; Dharmarajan, T.S.; Schierle, J.; Schalch, W. Serum lutein response is greater from free lutein than from esterified lutein during 4 weeks of supplementation in healthy adults. *J. Am. Coll. Nutr.* **2010**, *29*, 575–585. [CrossRef]
23. Wu, L.; Huang, X.; Shi, K.; Tan, R. Bioavailability comparison of free and esterified lutein for layer hens. *Braz. J. Poult. Sci.* **2009**, *11*, 95–98. [CrossRef]
24. Yoshizako, H.; Hara, K.; Takai, Y.; Kaidzu, S.; Obana, A.; Ohira, A. Comparison of macular pigment and serum lutein concentration changes between free lutein and lutein esters supplements in Japanese subjects. *Acta Ophthalmol.* **2016**, *94*, e411–e416. [CrossRef]
25. Moreiras, O.; Carbajal, A.; Cabrera, L.; Cuadrado, C. *Tablas de Composición de Alimentos*, 16th ed.; Ediciones Pirámide: Madrid, Spain, 2013.
26. Estévez Santiago, R.; Beltrán de Miguel, B.; Cuadrado Vives, C.; Olmedilla Alonso, B. Software application for the calculation of dietary intake of individual carotenoids and of its contribution to vitamin A intake. *Nutr. Hosp.* **2013**, *28*, 823–829. [PubMed]
27. Granado, F.; Olmedilla-Alonso, B.; Blanco, I.; Rojas-Hidalgo, E. Carotenoid composition in raw and cooked Spanish vegetables. *J. Agric. Food Chem.* **1992**, *40*, 2135–2140. [CrossRef]
28. Olmedilla, B.; Granado, F.; Blanco, I.; Rojas-Hidalgo, E. Quantitation of Provitamin-A and Non-Provitamin-A Carotenoids in the Fruits Most Commonly Consumed in Spain. In *Food and Cancer Prevention: Chemical and Biological Aspects*; Waldron, K.W., Johnson, I.T., Fenwick, G.R., Eds.; Royal Society of Chemistry: Cambridge, UK, 1993; pp. 141–145.
29. Beltrán, B.; Estévez, R.; Cuadrado, C.; Jiménez, S.; Olmedilla Alonso, B. Base de datos de carotenoides para valoración de la ingesta dietética de carotenos, xantofilas y de vitamina A: Utilización en un estudio comparativo del estado nutricional en vitamina A de adultos jóvenes. *Nutr. Hosp.* **2012**, *27*, 1334–1343. [PubMed]
30. Olmedilla, B.; Granado, F.; Gil-Martínez, E.; Blanco, I.; Rojas-Hidalgo, E. Reference levels of retinol, a-tocopherol and main carotenoids in serum of control and insulin-dependent diabetic Spanish subjects. *Clin. Chem.* **1997**, *43*, 1066–1071. [CrossRef]
31. Rodrigues, D.B.; Mercadante, A.Z.; Mariutti, L.R.B. Marigold carotenoids: Much more than lutein esters. *Food Res. Internat.* **2019**, *119*, 653–664. [CrossRef] [PubMed]
32. Eye Disease Case-Control Study Group. Antioxidant status and neovascular age-related macular degeneration. *Arch. Ophthalmol.* **1993**, *111*, 104–109. [CrossRef] [PubMed]
33. Granado, F.; Olmedilla, B.; Blanco, I. Nutritional and clinical relevance of lutein in human health. *Br. J. Nutr.* **2003**, *90*, 487–502. [CrossRef] [PubMed]
34. Ranard, K.M.; Jeon, S.; Mohn, E.S.; Griffiths, J.; Johnson, E.J.; Erdman, J.W., Jr. Dietary guidance for lutein: Consideration for intake recommendations is scientifically supported. *Eur. J. Nutr.* **2017**, *56* (Suppl. S3), 37–42. [CrossRef] [PubMed]
35. Eisenhauer, B.; Natoli, S.; Liewx, G.; Flood, M. Lutein and zeaxanthin- food sources, bioavailability and dietary variety in age-related macular degeneration protection. *Nutrients* **2017**, *9*, 120. [CrossRef]
36. Böhm, V.; Lietz, G.; Olmedilla-Alonso, B.; Phelan, D.; Reboul, E.; Bánati, D.; Borel, P.; Corte-Real, J.; de Lera, A.R.; Desmarchelier, C.; et al. From carotenoid intake to carotenoid blood and tissue concentrations—Implications for dietary intake recommendations. *Nutr. Rev.* **2021**, *79*, 544–573. [CrossRef]
37. Olmedilla-Alonso, B.; Beltran-de-Miguel, B.; Estevez-Santiago, R.; Cuadrado-Vives, C. Markers of lutein and zeaxanthin status in two age groups of men and women: Dietary intake, serum concentrations, lipid profile and macular pigment optical density. *Nutr. J.* **2014**, *13*, 52. [CrossRef]
38. Estévez-Santiago, R.; Beltrán-de-Miguel, B.; Olmedilla-Alonso, B. Assessment of dietary lutein, zeaxanthin and lycopene intakes and sources in the Spanish Survey of Dietary Intake (2009–2010). *Int. J. Food Sci. Nutr.* **2016**, *67*, 305–313. [CrossRef]
39. Bone, R.A.; Landrum, J.T. Dose-dependent response of serum lutein and macular pigment optical density to supplementation with lutein esters. *Arch. Biochem. Biophys.* **2010**, *504*, 50–55. [CrossRef] [PubMed]
40. Juturu, V.; Bowman, J.P.; Stringham, N.T.; Stringham, J.M. Bioavailability of lutein/zeaxanthin isomers and macular pigment optical density response to macular carotenoid supplementation: A randomized double blind placebo controlled study. *New Front. Ophtahlmol.* **2016**, *2*, 140–145. [CrossRef]
41. Machida, N.; Kosehira, M.; Kitaichi, N. Clinical effects of dietary supplementation of lutein with high bio-accessibility on macular pigment optical density and contrast sensitivity: A randomized double-blind placebo-controlled parallel-group comparison trial. *Nutrients* **2020**, *12*, 2966. [CrossRef] [PubMed]
42. Olmedilla, B.; Granado, F.; Southon, S.; Wright, A.J.A.; Blanco, I.; Gil-Martínez, E.; Van den Berg, H.; Corridan, B.; Hininger, I.; Thurnham, D.I.; et al. A European multicenter, placebo-controlled intervention trial with a-tocopherol, carotene rich palm-oil, lutein or lycopene at dietary achievable levels. *Clin. Sci.* **2002**, *102*, 447–456. [CrossRef]
43. Breithaupt, D.E.; Weller, P.; Wolters, M.; Hahn, A. Comparison of plasma responses in human subjects after the ingestion of 3R,3R′-zeaxanthin dipalmitate from wolfberry (*Lycium barbarum*) and non-esterified 3R,3R′-zeaxanthin using chiral high-performance liquid chromatography. *Br. J. Nutr.* **2004**, *91*, 707–713. [CrossRef]
44. Stringham, J.M.; Hammond, B.R., Jr. The glare hypothesis of macular pigment function. *Optom. Vis. Sci.* **2007**, *84*, 859–864. [CrossRef]

45. Hammond, B.R.; Fletcher, L.M.; Roos, F.; Wittwer, J.; Schalch, W. A double-blind, placebo-controlled study on the effects of lutein and zeaxanthin on photostress recovery, glare disability, and chromatic contrast. *Investig. Ophthalmol. Vis. Sci.* **2014**, *55*, 8583–8589. [CrossRef] [PubMed]
46. Liu, R.; Wang, T.; Zhang, B.; Qin, L.; Wu, C.; Li, Q.; Ma, L. Lutein and zeaxanthin supplementation and association with visual function in age-related macular degeneration. *Investig. Ophthalmol. Vis. Sci.* **2014**, *56*, 252–258. [CrossRef]
47. Estévez-Santiago, R.; Olmedilla-Alonso, B.; Beltrán-de-Miguel, B. Assessment of lutein and zeaxanthin status and dietary markers as predictors of the contrast threshold in 2 age groups of men and women. *Nutr. Res.* **2016**, *36*, 719–730. [CrossRef] [PubMed]
48. Lawler, T.; Liu, Z.; Nalbandyan, M.; Liu, Y.; Hammond, B.; Wallace, R.B.; Mares, J.A. Lutein and zeaxanthin supplement use is associated with increased macular pigment density over 15 years and greater contrast sensitivity in the Carotenoids in Age-Related Eye Disease Study of older-adult women. *Investig. Ophthalmol. Vis. Sci.* **2021**, *62*, 2950.
49. Yao, Y.; Qiu, A.-H.; Wu, S.-W.; Cai, Z.-Y.; Xu, S.; Liang, X.-Q. Lutein supplementation improves visual performance in Chinese drivers: 1-year randomized, double-blind, placebo-controlled study. *Nutrition* **2013**, *29*, 958–964. [CrossRef] [PubMed]
50. Ma, L.; Lin, X.-M.; Zou, A.-Y.; Xu, X.-R.; Li, Y.; Xu, R. A 12-week lutein supplementation improves visual function in Chinese people with long-term computer display light exposure. *Br. J. Nutr.* **2009**, *102*, 186–190. [CrossRef] [PubMed]
51. Ma, L.; Yan, S.-F.; Huang, Y.-H.; Lu, X.-R.; Qian, F.; Pan, H.-L.; Xu, X.R.; Zou, Z.Y.; Dong, P.C.; Xiao, X.; et al. Effect of lutein and zeaxanthin on mcular pigment and visual function in patients with early age-related macular degeneration. *Ophthalmology* **2012**, *119*, 2290–2297. [CrossRef] [PubMed]
52. Du, M.; Wu, M.; Fu, D.; Yang, S.; Chen, J.; Wilson, K.; Lyons, T.J. Effects of modified LDL and HDL on retinal pigment epithelial cells: A role in diabetic retinopathy? *Diabetologia* **2013**, *56*, 2318–2328. [CrossRef]
53. Fernandez-Garcia, E.; Carvajal-Lerida, I.; Jaren-Galan, M.; Garrido-Fernandez, J.; Perez-Galvez, A.; Hornero-Mendez, D. Carotenoids bioavailability from foods: From plant pigments to efficient biological activities. *Food. Res. Int.* **2012**, *46*, 438–450. [CrossRef]

nutrients

MDPI

Article

Purple Corn Extract Improves Dry Eye Symptoms in Models Induced by Desiccating Stress and Extraorbital Lacrimal Gland Excision

Jae-Min Lee, Arin Choi, Hee-Hwan Lee, Sang Jae Park and Byung-Hak Kim *

MEDIENCE Co., Ltd., Chuncheon 24232, Republic of Korea; jmlee@mdec.co.kr (J.-M.L.); archoi@mdec.co.kr (A.C.); hhlee@mdec.co.kr (H.-H.L.); psmarch@naver.com (S.J.P.)

* Correspondence: bhkim@mdec.co.kr; Tel.: +82-33-258-6256

Abstract: Dry eye disease (DED) occurs when there are not enough tears, and the associated symptoms—burns, itching, and a gritty feeling in the eye—can cause great discomfort. The purpose of this study was to evaluate the therapeutic effect of purple corn extract (PCE) on DED. Pretreatment with PCE prevented desiccation-stress-induced cell damage in human retinal pigment epithelial cells and primary human corneal epithelial cells. Furthermore, PCE reduced the mRNA expression of inflammatory mediators in the induction of desiccation stress. The therapeutic effects of PCE on DED were evaluated in an animal model with induced unilateral excision of the exorbital lacrimal gland. The administration of PCE was effective at recovering tear production, corneal surface irregularity, and conjunctival goblet cell density, as well as at reducing apoptotic cell death in the outer layer of the corneal epithelium. Collectively, PCE improved dry eye symptoms, and, therefore, it could be a potential agent to ameliorate and/or treat DED.

Keywords: dry eye disease; purple corn extract; anthocyanin; cyanidin-3-*O*-glucoside; inflammation; regulated cell death

1. Introduction

Age-related dry eye disease (DED) is a multifactorial and common ocular surface disorder causing immense inconvenience, which adversely affects the quality of life of patients. With increasing longevity, this is a growing health problem globally, which affects vision in severe cases [1–3]. DED is primarily classified in two categories: aqueous-deficient DED with lack of tear production and evaporative DED with increased tear evaporation [1,4]. The ocular surface plays an essential role in maintaining moisture via the tear film, which is composed of several components, including water, lipids, mucins, cytokines, growth factors, and antimicrobial peptides [5,6]. On the ocular surface, the tear film lubricates and defends against pathogens. As we age, the tear film homeostasis changes, affecting the lacrimal glands, goblet cells, and meibomian glands in particular. These changes may lead to aqueous-deficient DED and further injury to the ocular surface through decreased secretion of tear volume [7–10].

Patients with DED may experience many uncomfortable symptoms, including photophobia, pain, burns, itching, and a gritty feeling in the eye. Although aging is a gradual and continuous process of natural change, it can be accompanied by numerous biological and physiological changes. Inflammaging, which is a chronic and low-grade inflammatory state, occurs during the aging process. It is characterized by an increase in the production of pro-inflammatory mediators. This condition accelerates tear instability, with tear production and DED symptoms, and the induction of stress-related signal transductions on the ocular surface [3,11,12]. Consequently, the regulation of inflammation can be a favorable therapeutic approach for DED. In fact, anti-inflammatory agents are used to reduce the signs and symptoms of DED and restore homeostasis in the tear film and ocular surface [12–14].

Originally, purple corn (*Zea mays* L.) came from the Andean Region of South America and was used to make traditional foods and beverages. It is now widely cultivated all over the world and has recently become an emerging star in the novel ingredients market and the pharmaceutical industry, as it includes a number of biologically active compounds, such as anthocyanins and other functional phytochemicals, that are beneficial for health [15,16]. Among the anthocyanins, cyanidin-3-*O*-glucoside (C3G), the 3-*O*-glycosylated form of cyanidin, is one of the most common rich compounds of purple corn. It has been reported to have several health-promoting effects, such as antioxidant, anti-inflammation, antidiabetic, antiangiogenetic, and anticarcinogenic properties [15–18]. Because of their high anthocyanin content, the husks and cobs of purple corn can be optional sources of anthocyanin for food supplements; however, they are commonly recognized as by-products and are largely discarded because of their low utilization. For this reason, it is essential to develop a number of products with benefits for health that can be obtained from the husks and cobs of purple corn. In the present study, we prepared purple corn extract (PCE) from the husks and cobs of purple corn and evaluated the biological activities of PCE on eye health, especially DED, in both in vitro and in vivo models.

2. Materials and Methods

2.1. Preparation of Purple Corn Extract

The hybrid grains of purple corn were developed and registered by Gangwon Agricultural Research and Extension Services (Chuncheon, Republic of Korea). As previously described, the PCE was prepared from the dried husks and cobs of the purple corn [19]. Briefly, equal amounts of husks and cobs were repeatedly extracted with a 10-fold volume of 30% ethanol for 6–12 h at 60 °C. The extract was subsequently filtered using Whatman Qualitative No. 1 filter paper, concentrated at 60 °C in a rotary evaporator (Eyela, Tokyo, Japan), and PCE powder was obtained by spray drying using a spray drier. The powder was dissolved in dimethyl sulfoxide (DMSO) as a stock solution at a concentration of 200 mg/mL, and in vitro biological assays were then carried out. The DMSO used as a vehicle had a final concentration of 0.15%, except for the cell viability assay.

2.2. High-Performance Liquid Chromatography (HPLC) Analysis of Cyanidin-3-O-Glucoside

Cyanidin-3-*O*-glucoside (C3G) is one of the major components of PCE. In order to quantify the amounts of C3G, the extracted PCE was analyzed using an LC-20AD pump (Kyoto, Japan). Quantitative analysis was carried out using a reverse-phase system, and separation was performed on a Unison US-C18 column (25 cm $\times$ 4.6 mm, 5 μm). The column temperature was maintained at 40 °C, and the flow rate was set to 1 mL/min. The mobile phase comprised 0.1% trifluoroacetic acid (TFA) in water and acetonitrile, and the elution was carried out at 85–15% for 40 min. The injection volume was 10 μL, and the UV absorbance was recorded at 525 nm. The standard reagent against C3G was obtained from ChromaDex (Los Angeles, CA, USA).

2.3. Cell Culture

The human retinal pigment epithelial cell line ARPE-19 (CRL-2302) and primary human corneal epithelial cells (HCEpiC, PCS-700-010) were obtained from the American Type Culture Collection (Manassas, VA, USA). The ARPE-19 cells were maintained in DMEM/F-12 (HyClone, Logan, UT, USA) and supplemented with 10% heat-inactivated fetal bovine serum (FBS, HyClone) and 1% penicillin/streptomycin (Welgene, Gyeongsan, Republic of Korea). According to the manufacturer's instructions, primary HCEpiCs were maintained in corneal epithelial cell basal medium (PCS-700-030, ATCC) that contained growth supplements (PCS-700-040, ATCC).

2.4. Cell Viability Assay

Cells were seeded at a density of 2×10^4 cells/well in 96-well plates containing fresh medium. When the cells reached their experimental time points after an overnight incubation, the cells were incubated with vehicle alone, with different concentrations of PCE, or with omega-3 for 24–72 h. The cell viability was determined at 450 nm using a microplate reader (Molecular Devices, Sunnyvale, CA, USA) after being further incubated for 2–4 h at 37 °C following the addition of EZ-CyTox Enhanced Cell Viability Assay Reagent (Daeil Lab Service, Seoul, Republic of Korea). In order to induce desiccating stress, the cells were exposed at various time points of the desiccation stress or exposed to desiccation stress for 30 min, followed by the removal of the cultured medium, and then the cell viability was determined using an EZ-CyTox Enhanced Cell Viability Assay Reagent. The crude source of omega-3 (F8020) used as a positive control was obtained from Sigma-Aldrich (St. Louis, MO, USA).

2.5. Quantitative Real-Time Polymerase Chain Reaction (qRT-PCR)

According to the manufacturer's instructions, the total RNA was extracted using an AccuPrep Universal RNA Extraction Kit (Bioneer, Daejon, Republic of Korea), and the cDNA was synthesized using an WizScript cDNA Synthesis Kit (Wisbio Solutions, Seongnam, Republic of Korea). Using a Dyne Fast qPCR 2× PreMix (Dyne Bio, Seongnam, Republic of Korea) with the Mic qPCR Analysis Software v2.12.0 (Bio Molecular Systems, Upper Coomera, Australia), the qRT-PCR was performed. A comparative C_t quantification of the data was performed to derive a value for each sample, which was normalized to the value for the housekeeping glyceraldehyde-3-phosphate dehydrogenase (GAPDH) gene. The sequences of the primers are summarized in Table 1.

Table 1. Primer sequences used for the quantitative real-time PCR.

Target Genes		Primer Sequences
IL-1β	Forward	5′-GCACGATGCACCTGTACGAT-3′
	Reverse	5′-AGACATCACCAAGCTTTTTTGCT-3′
IL-6	Forward	5′-CAGGAATTGAATGGGTTTGC-3′
	Reverse	5′-AAACCAAGGCACAGTGGAAC-3′
IL-8	Forward	5′-TTTTGCCAAGGAGTGCTAAAGA-3′
	Reverse	5′-AACCCTCTGCACCCAGTTTTC-3′
IL-12	Forward	5′-CTTGTGGCTACCCTGGTCCT-3′
	Reverse	5′-GAGTTTGTCTGGCCTTCTGG-3′
TNF-α	Forward	5′-CTGGGCAGGTCTACTTTGGG-3′
	Reverse	5′-CTGGAGGCCCCAGTTTGAAT-3′
iNOS	Forward	5′-GGTGGAAGCGGTAACAAAGG-3′
	Reverse	5′-TGCTTGGTGGCGAAGATGA-3′
COX-2	Forward	5′-GCCAAGCACTTTTGGTGGAG-3′
	Reverse	5′-GGGACAGCCCTTCACGTTAT-3′
GAPDH	Forward	5′-GACCACAGTCCATGCCATCA-3′
	Reverse	5′-TCCACCACCCTGTTGCTGTA-3′

2.6. Animal Experiments

Six-week-old male Sprague–Dawley (SD) rats were obtained from KoaTech (Pyeongtaek, Republic of Korea), and these rats were acclimatized for 7 days. In order to induce experimental DED, the rats were deeply anesthetized with isoflurane (JW Pharmaceutical, Seoul, Republic of Korea), and their left exorbital lacrimal gland was surgically excised. The next day, the rats were randomly separated into five groups ($n = 5$) and orally administered either the vehicle (vehicle-treated group), PCE (30, 80, or 150 mg/kg, PCE-treated group), or omega-3 (100 mg/kg, omega-3-treated group) every day for 7 days. Sham-operated rats were assigned to the normal group. Prior to administration, the PCE and omega-3 were freshly prepared in distilled water every day. The experimental protocol was approved by the Institutional Animal Care and Use Committee (IACUC) of Chonbuk National University Hospital Non-Clinical Evaluation Center (IACUC approval No. 2022-42).

2.7. Tear Volume Measurement

Tear volume was assessed using the phenol red thread tear test (Zone Quick, FCI Ophthalmics, Pembroke, MA, USA) following treatment with PCE or omega-3 for 7 days. The cotton thread was placed in the lateral canthus for 30 s, and the length of the color-changed thread was measured. The tear volume was measured in both eyes.

2.8. Corneal Surface Irregularity

As previously mentioned, the change in the ocular surface induced by desiccating stress was determined [20]. The irregularity was measured by the reflected light from the fiber-optic ring illuminator (SZ51, Olympus, Tokyo, Japan) on the corneal surface. The values were scored as follows: 0, no distortion; 1, distortion in 1/4 of the reflected ring shape; 2, distortion in 2/4; 3, distortion in 3/4; 4, distortion in 4/4; and 5, severe distortion and no ring shape could be recognized.

2.9. Periodic Acid–Schiff (PAS) Staining

Five-micrometer-thick sections from the paraffin-embedded cornea specimens were mounted on glass slides, deparaffinized, hydrated with grade ethanol, and then oxidized in a 1% periodic acid solution for 5–10 min. These sections were rinsed with distilled water and reacted with Schiff's reagent for 20–30 min. After being rinsed with tap water, the sections were dehydrated with grade ethanol, cleared with xylene, mounted in DPX (Sigma-Aldrich), and then observed under an inverted fluorescence microscope (Carl Zeiss, Jena, Germany). Counterstaining of the sections was performed using 0.2% Mayer's Hematoxylin Solution (Sigma-Aldrich).

2.10. Terminal Deoxynucleotidyl Transferase-Mediated dUTP Nick End Labeling (TUNEL) Staining

According to the manufacturer's instructions (In Situ Cell Death Detection Kit, Roche, Germany), the apoptotic cells on the corneal tissue were determined with TUNEL staining. The TUNEL-positive cells were counted using an inverted fluorescence microscope (Carl Zeiss).

2.11. Statistical Analysis

The results were derived from at least three independent experiments, and all data are presented as the mean ± standard error of mean (SEM). The statistical analysis was carried out with the use of GraphPad Prism 5.0 (GraphPad Software, San Diego, CA, USA), and the significance was determined using a two-tailed Student's *t*-test. Differences were considered statistically significant at $p < 0.05$.

3. Results

3.1. HPLC Analysis of PCE

Purple corn contains a wide range of anthocyanins that are present in its husks, barks, stems, and kernels, with C3G being one of the most common constituents and responsible for its purple color [16–18,21,22]. Developed by Gangwon Agricultural Research and Extension Services, the PCE was prepared from a mixture of the dried husks and cobs of the purple corn. We obtained a standard compound of C3G and performed an HPLC analysis to determine the amount of C3G in the PCE. The data from the HPLC analyses of the C3G standard and the PCE used in the chromatograms were obtained by observing the detector responses at 525 nm. We observed two major peaks in the chromatogram of the

PCE, one of which, C3G, was detected at about 7.8 min; its content was 18.43 ± 0.58 mg/g (Figure 1).

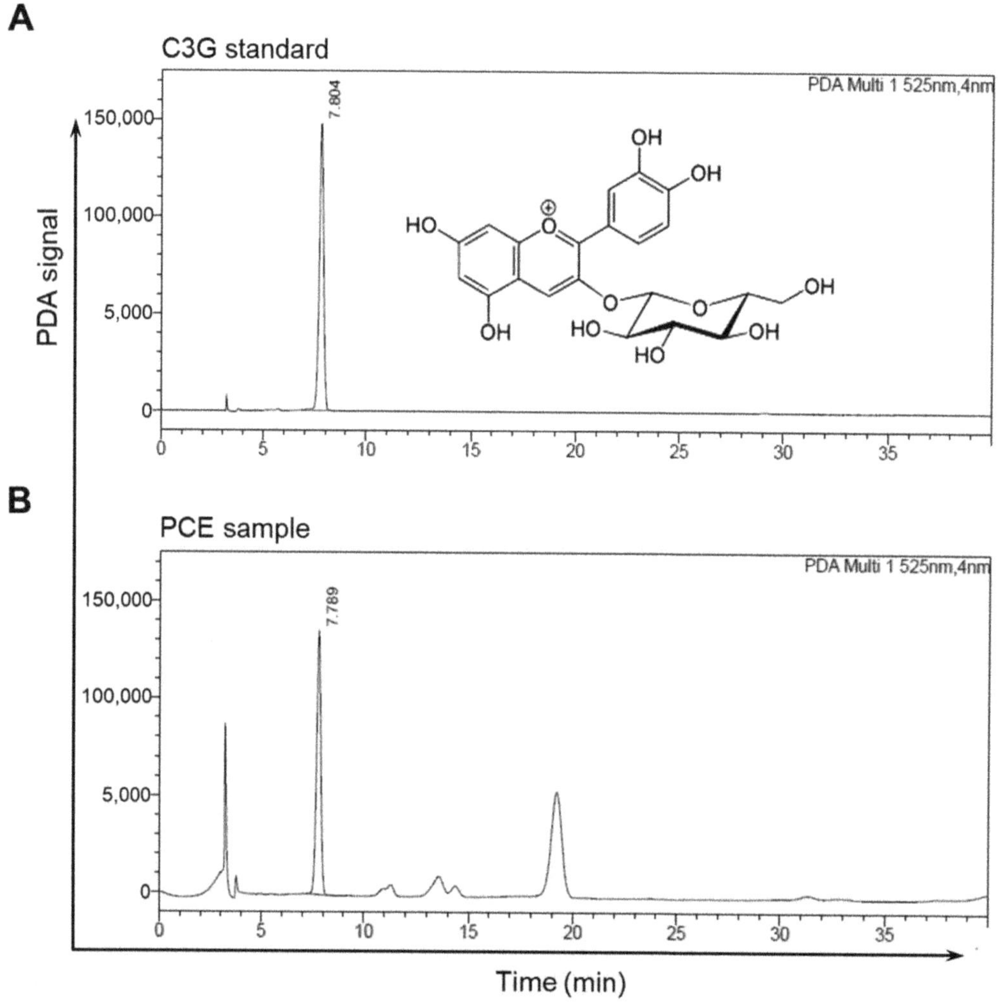

Figure 1. Representative chromatographs of HPLC: (**A**,**B**) HPLC chromatograms of the C3G standard (**A**) and PCE (**B**) were detected at 525 nm. The peak of C3G appeared at a retention time of 7.8 min.

3.2. PCE Protects Cells against Damage from Desiccation Stress

The cell viability assay was performed in human retinal pigment epithelial and primary human corneal epithelial cells to determine the cytotoxicity of PCE. The cells were maintained for up to 72 h with various concentrations of PCE or omega-3. The PCE did not affect any cytotoxic activity up to 3000 μg/mL for 72 h in both cell lines (Figure 2A,B). However, in the cell lines with high concentrations above 1000 μg/mL, omega-3 enhanced the viability of cells (Figure 2C,D).

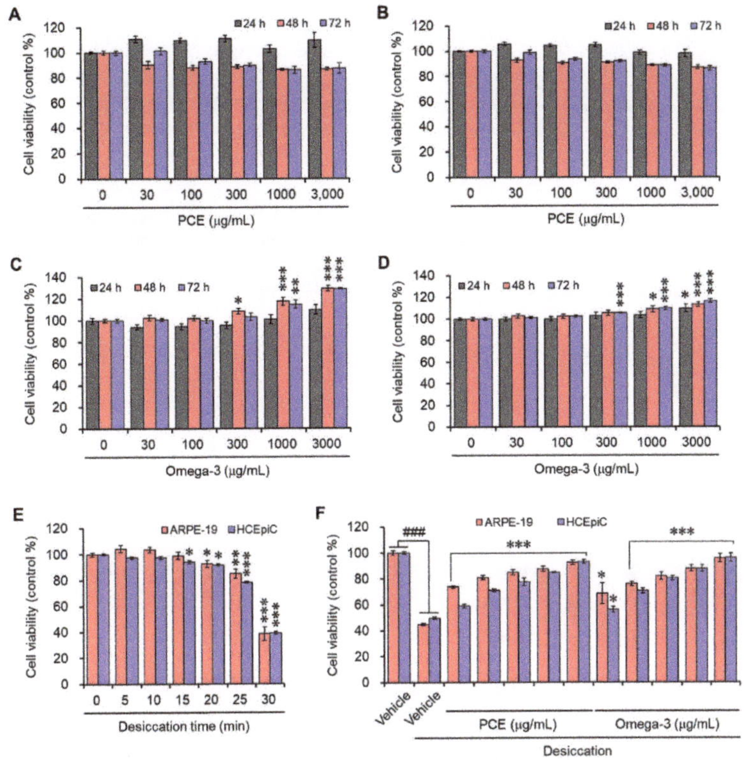

Figure 2. Effects of PCE and omega-3 on the viability of ARPE-19 and HCEpiCs: (**A–D**) ARPE-19 (**A**,**C**) or HCEpiCs (**B**,**D**) were incubated with different concentrations of C3G (**A**,**B**) or omega-3 (**C**,**D**) for the indicated time points, and their viability was determined. (**E**) ARPE-19 or HCEpiCs were exposed to different time points of air to induce desiccation stress, and their viability was determined. In all cases, the viability is expressed as a % of the control. The results are presented as the mean ± SEM ($n = 3$). * $p < 0.05$, ** $p < 0.005$, and *** $p < 0.0001$ versus the vehicle-treated group. (**F**) ARPE-19 or HCEpiCs were preincubated with different concentrations of C3G or omega-3 for 24 h; then, they were exposed to desiccation stress for 30 min, and the cell viability was determined. In all cases, the viability is expressed as a % of the control. The results are presented as the mean ± SEM ($n = 3$). ### $p < 0.0001$ versus the vehicle-treated group. * $p < 0.05$ and *** $p < 0.0001$ versus the desiccation-stress-induced vehicle group.

Desiccation stress was induced in ARPE-19 and HCEpiC cells at different time points to mimic DED in a cell culture system, and then the cell viability was observed. As previously reported, desiccation stress was induced through air exposure on a clean bench [23]. The cell viability slightly decreased at 20 min and dramatically decreased at 30 min (Figure 2E). Therefore, the time point for desiccation stress was set to 30 min for the in vitro experiments. To examine the protective effect of the PCE or omega-3 against cell damage induced by desiccation stress, ARPE-19 or HCEpiC cells were preincubated with different concentrations of PCE or omega-3 for 24 h prior to exposure to desiccation stress for 30 min. In the presence of PCE or omega-3, the cell viability was restored effectively in a concentration-dependent manner, and this indicates the protective effects of PCE or omega-3 against desiccation stress (Figure 2F). On the basis of the results of the cell viability assays, we used concentrations of up to 300 µg/mL of PCE or omega-3 in the in vitro experiments.

3.3. PCE Decreases the mRNA Level of Inflammatory Mediators

Inflammation is known to further accelerate DED and is an important DED biomarker that is considered in the regulated cell death (RCD) of dry eye etiopathology [24]. Interleukin-6 (IL-6) and tumor necrosis factor-alpha (TNF-α), in particular, are correlated with necroptosis, and IL-1β is associated with pyroptosis [24]. To confirm whether the cellular cytoprotective effect of PCE is due to its anti-inflammatory effect, the mRNA expression of inflammatory mediators was determined with qRT-PCR. The mRNA expression of inflammatory molecules, which include IL-1β, IL-6, IL-8, IL-12, TNF-α, inducible nitric oxide synthase (iNOS), and cyclooxygenase-2 (COX-2), was increased approximately 2–4-fold that of the normal conditions via the induction of desiccation stress in the ARPE-19 and HCEpiC cells. This demonstrates that under desiccation conditions, the inflammatory response is activated. In a concentration-dependent manner, the PCE caused an effective decrease in their mRNA expression, and its effects were comparable to or stronger than those of omega-3 (Figures 3 and 4). These findings demonstrate that PCE has anti-inflammatory effects in dry eye syndromes, and this activity may play an essential role in its cytoprotective effect.

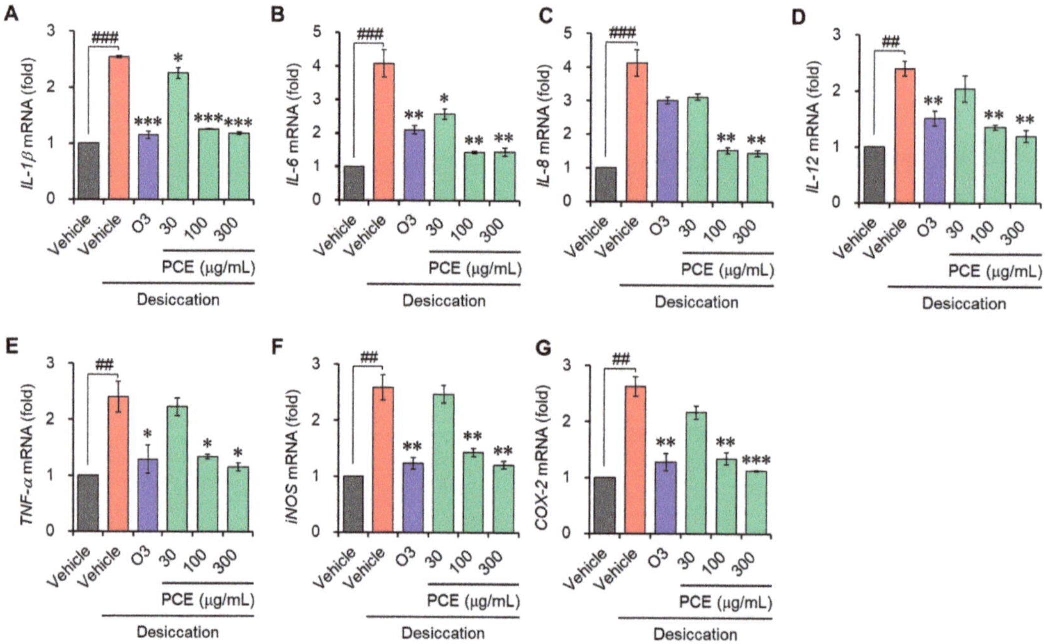

Figure 3. Effects of PCE on mRNA expression of inflammatory mediators in ARPE-19: (**A–G**) ARPE-19 cells were preincubated with different concentrations of PCE for 2 h, exposed to desiccation stress for 30 min, and further incubated for 6 h. The total RNA was extracted, and the mRNA expression of inflammatory mediators was analyzed using real-time PCR. The results are expressed as relative values compared to the vehicle-treated control and presented as the mean ± SEM (n = 3). ## $p < 0.005$ and ### $p < 0.0001$ versus the vehicle-treated group. * $p < 0.05$, ** $p < 0.005$, and *** $p < 0.0001$ versus the desiccation-stress-induced vehicle group. O3, omega-3, 300 μg/mL.

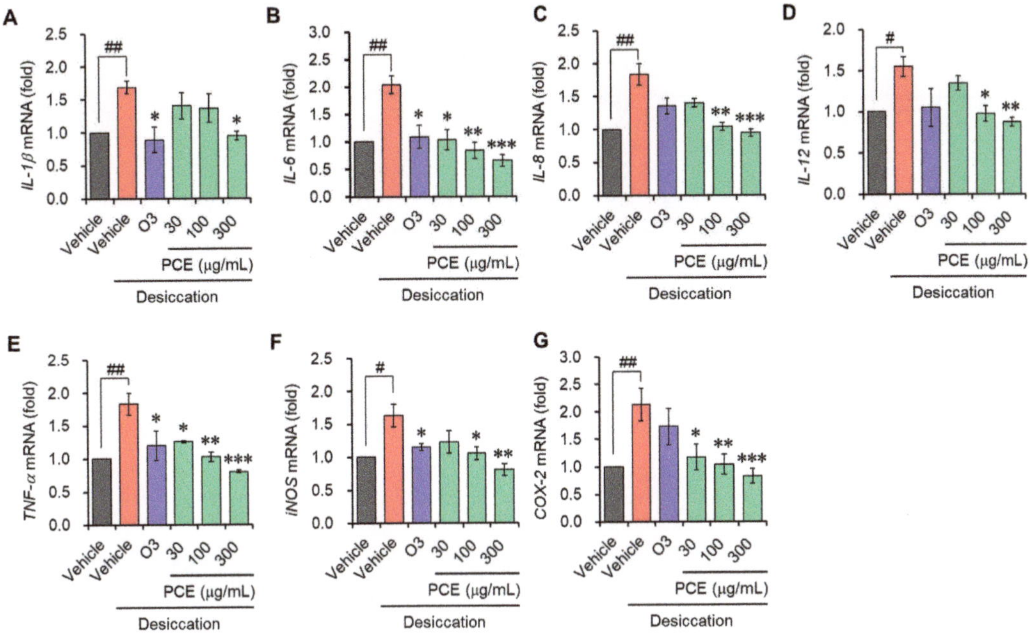

Figure 4. Effects of PCE on the mRNA expression of inflammatory mediators in HCEpiCs: (**A–G**) HCEpiCs were preincubated for 2 h with different concentrations of PCE, exposed to desiccation stress for 30 min, and further incubated for 6 h. The total RNA was extracted, and the mRNA expression of the inflammatory mediators was analyzed using real-time PCR. The results are expressed as relative values compared to the vehicle-treated control and presented as the mean ± SEM ($n = 3$). # $p < 0.05$ and ## $p < 0.005$ versus the vehicle-treated group. * $p < 0.05$, ** $p < 0.005$, and *** $p < 0.0001$ versus the desiccation-stress-induced vehicle group. O3, omega-3, 300 μg/mL.

3.4. PCE Improves Tear Production and Corneal Irregularity in Rats with DED

An in vivo model of DED was developed by surgical excision of the unilateral exorbital lacrimal gland in SD rats to evaluate the effect of PCE on eye health. PCE was administered orally at a constant concentration every day for 7 days, and then the tear volume was measured using a phenol-red-impregnated cotton thread. Tear production was reduced compared with the normal group for rats that were experimentally exposed to a DED model, suggesting that the DED model was well established. PCE administration effectively recovered the tear volume in a concentration-dependent manner. At high concentrations, the effect was almost normal and stronger than that of omega-3 (Figure 5A).

DED is associated with increased frequent blinking without the production of tears, causing damage to the cornea surface due to scratching. These phenomena alter the corneal epithelial barrier function and tear film integrity and cause inflammation. Therefore, the extent of corneal damage was determined by taking pictures of the rat's eye, which reflected a white ring of light. The DED model group exhibited irregular circle shapes of light; however, the administration of PCE significantly restored these shapes to those of the normal group, and the effect was stronger than that of omega-3 (Figure 5B,C). These findings suggest that PCE may possibly improve eye health, especially DED, by maintaining moisture in the eye.

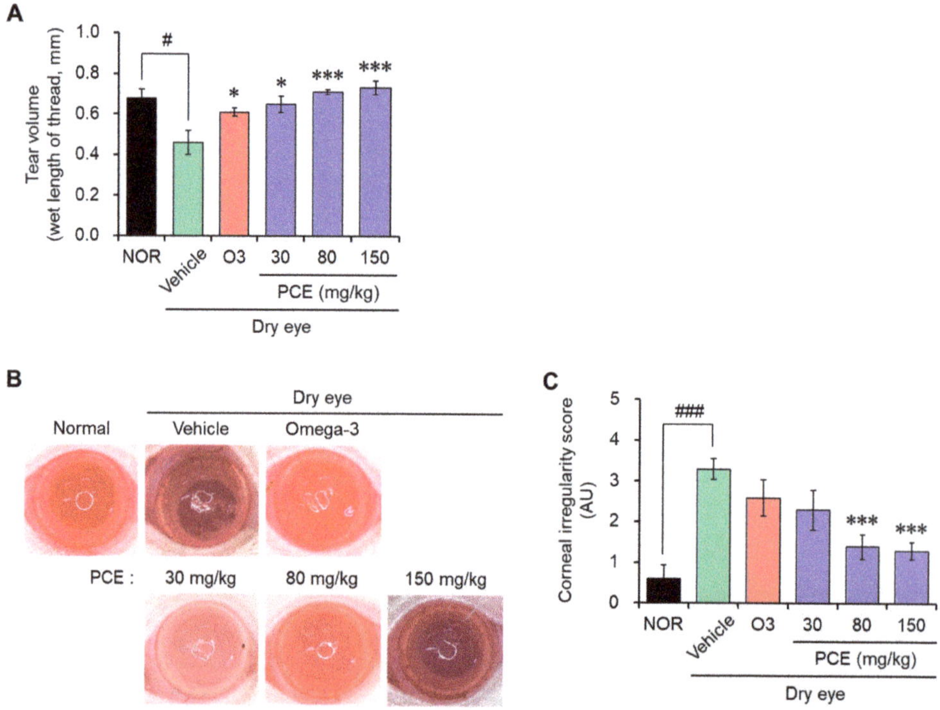

Figure 5. Effects of PCE on tear secretion and corneal irregularity in DED model rats. (**A**) Tear volume was measured using a phenol red thread test. Tear volume is expressed in milliliters of thread wetted by the tear and turned red in color. (**B**,**C**) Reflected images (**B**) and the scores (**C**) of a white ring from the fiber-optic ring illuminator of a stereomicroscope, and the corneal irregularity was graded based on the number of distorted quarters in the reflected white ring, as described in Section 2.8. The results are presented as the mean ± SEM ($n = 5$). $^{\#}$ $p < 0.05$ and $^{\#\#\#}$ $p < 0.0001$ versus the normal group. * $p < 0.05$ and *** $p < 0.0001$ versus the DED-induced group. O3, omega-3, 100 mg/kg.

3.5. PCE Increases Corneal Goblet Cell Density on the Ocular Surface

Corneal mucins are essential to defend and preserve the integrity of the ocular surface, and alterations in corneal mucins have been found to be hallmarks of DED. Specifically, a decrease in goblet cell density is associated with the seriousness of DED, and this decrease is accompanied by a decrease in mucin production [25,26]. Therefore, the density of the goblet cells was analyzed for an indirect assessment of mucin production using immunohistochemistry for PAS staining, which can selectively stain acidic mucus, a mucin component contained in goblet cells, with red color. The goblet cell density remarkably decreased in the rats in the DED model group compared to those in the normal group, but it effectively increased in a concentration-dependent manner with the administration of PCE, and its effect was comparable to or stronger than that of omega-3 (Figure 6). The protective effect of PCE on the goblet cell density clearly indicates that PCE probably protects corneal surface damage and maintains tear film integrity by increasing the production of mucin, thereby improving dry eye symptoms.

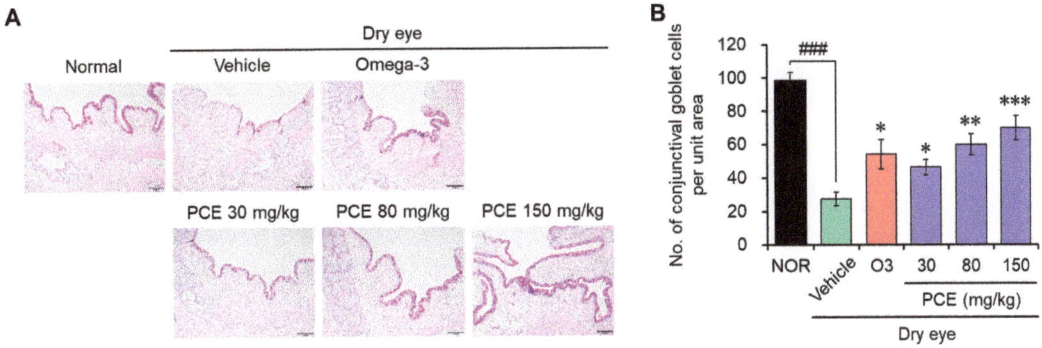

Figure 6. Effects of PCE on goblet cell density in DED model rats: (**A**,**B**) immunohistochemical analysis of the PAS staining was carried out to analyze the acidic mucus in goblet cells, and representative images (**A**) and PAS-positive cells in the corneal sections are shown as the conjunctival goblet cell density (**B**). The results are presented as the mean ± SEM ($n = 5$). ### $p < 0.0001$ versus the normal group. * $p < 0.05$, ** $p < 0.005$, and *** $p < 0.0001$ versus the DED-induced group. O3, omega-3, 100 mg/kg.

3.6. PCE Prevents Apoptotic Cell Death in the Corneal Epithelium

DED is mainly linked to a lack of moisture in the eye, which leads to inflammation resulting in cell death associated with stress [24,27]. Recently, targeting RCD has been an emerging field and innovative strategy in the therapeutics of DED and ocular surface dysfunction. Therefore, protecting cell death from cell damage caused by stress is an important modulator of DED-related eye and ocular health improvement. In order to determine whether PCE has cytoprotective effects towards an in vivo model of DED, TUNEL staining was conducted in the tissue sections of the cornea of rats to detect dead cells by mediated apoptosis. TUNEL-positive cells were dramatically increased in the corneal epithelium and lacrimal glands of the DED rats compared to those in the normal control rats. This indicates that RCD, particularly apoptotic cell death, increased. However, this increase in apoptotic cell death was effectively reduced by PCE administration, and this effect was comparable to or stronger than that of omega-3 (Figure 7).

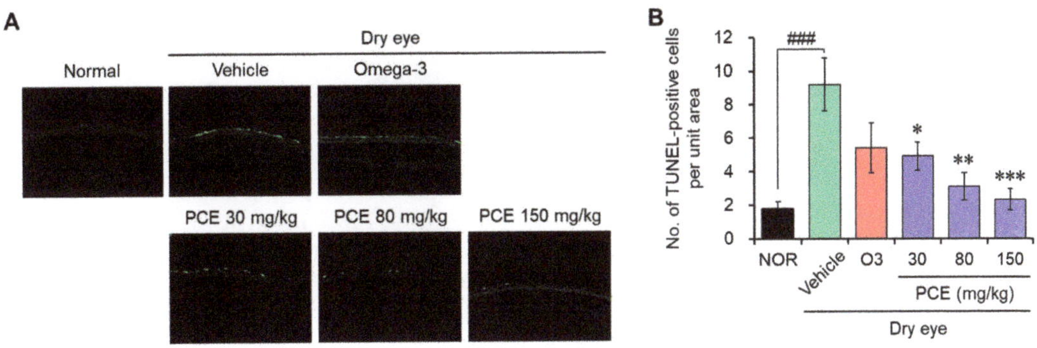

Figure 7. Effects of PCE on apoptotic cell death in the corneal epithelium of DED model rats: (**A**,**B**) TUNEL staining was carried out to analyze apoptotic cell death in the corneal epithelium, and representative images (**A**) and TUNEL-positive cells in the corneal sections (**B**) are shown. The results are presented as the mean ± SEM ($n = 5$). ### $p < 0.0001$ versus the normal group. * $p < 0.05$, ** $p < 0.005$, and *** $p < 0.0001$ versus the DED-induced group. O3, omega-3, 100 mg/kg.

4. Discussion

DED, one of the most prevalent eye syndromes, affects millions of people around the world and becomes more common as people age. Recently, the incidence rate has rapidly been increasing in the younger population due to frequent use of mobile phones and video screens, use of contact lenses, and environmental factors, such as air pollution and low humidity [2,28]. These factors decrease tears in the eyes, leading to hyperosmolarity of tear film and inflammation of the ocular surface that causes damage to the ocular surface. Tear production deficiency is associated with various pathophysiological mechanisms such as decreasing cell volume, dysfunction of the DNA repair systems, and promotion of reactive oxygen species (ROS) generation and apoptotic cell death in ocular surface cells [20,29]. Therefore, care and attention should be paid to eye health, not only for middle-aged people but also for early age people.

Artificial tears, lubricants, and various ophthalmic drops, such as anti-inflammatory drugs, immunosuppressants, and steroids, are currently used to treat DED, but these therapeutics concentrate on relieving clinical symptoms rather than unraveling the fundamental causes of DED [30]. These limitations indicate the need to develop novel therapeutic agents to treat DED. We investigated the protective effects of purple corn on DED models for the improvement of eye health by treating the fundamental causes of DED. Purple corn is a natural product, which is newly developed and registered by Gangwon Agricultural Research and Extension Services, and the extract, PCE, was prepared from a mixture of the husks and cobs of purple corn [19]. To examine the potential protective effects of PCE against DED, the DED models were constructed with desiccation stress induced by air exposure on a clean bench in vitro and by surgical excision of the unilateral exorbital lacrimal gland in SD rats.

Purple corn contains various biologically functional components such as anthocyanins, phenolic acids, and flavonoids. These components play an important role in its health benefits, which include antioxidant, anti-inflammatory, antimutagenic, anticarcinogenic, anticancer, and antiangiogenetic properties; blood pressure regulation; cardiovascular health benefits; and improvement in lifestyle diseases, like obesity, diabetes, hyperglycemia, and other linked diseases [15–18]. On the basis of our findings, PCE effectively improved eye health in both the in vitro and in vivo DED models. PCE exhibited cytoprotective and anti-inflammatory activities in ARPE-19 and HCEpiC cells induced by desiccation stress in the in vitro experiments. Specifically, the anti-inflammatory activity of PCE may be closely associated with cytoprotective effects, because inflammatory molecules promote RCD in DED conditions during inflammaging [3,11,12,24]. Particularly, inflammatory cytokines, such as IL-1β, IL-6, IL-17A, IL-18, and TNF-α, are associated with RCDs, such as apoptosis, necroptosis, and pyroptosis [24]. Using TUNEL staining, the cytoprotective effect of PCE was further validated in the in vivo model of DED. PCE administered orally significantly reduced TUNEL-positive cells in the corneal epithelium and lacrimal glands of the DED rats, which indicates the anti-apoptotic activity of PCE. It is thus predicted that the cytoprotective effects of PCE have a strong correlation with its anti-inflammatory activity.

Anthocyanins, which are a class of water-soluble phenolic compounds with a content in the range of 6.8–82.3 mg/g, are mainly responsible for the dark purple–red color of purple corn. The anthocyanin content in purple corn is much higher than that in other products known to be rich in anthocyanin, such as red grapes, blueberries, and aronia [21,31,32]. The major anthocyanins in purple corn are C3G, pelargonidin-3-*O*-glucoside, peonidin-3-*O*-glucoside, cyanindin-3-*O*-(6″-malonylglucoside), pelargonidin-3-*O*-(6″-malonylglucoside), and peonidin-3-*O*-(6″-malonylglucoside) [21,33–37]. It is well known that these components play an essential role in beneficial biological activities. In the HPLC analysis of the PCE, two major peaks and three minor peaks were observed; one of the major peaks was identified as C3G, and its content was 18.43 $\pm$ 0.58 mg/g. Interestingly, Wistar rats and F344 rats in acute and 90-day subchronic oral toxicity studies showed no toxicity symptoms at high dosages of purple corn pigment, suggesting that the products made by purple corn are safe [38,39]. According to the results reported by Nabae and colleagues, the no-observed-adverse-effect level (NOAEL) values were determined to be 3542 and 3849 mg/kg body weight/day in male and female F344 rats, respectively [39]. To date, PCE has been found to be safe enough that there is no limitation to the recommended dosage of the consumption of purple corn anthocyanin for human health.

According to the Tear Film & Ocular Surface Society (TFOS) Dry Eye Workshop II (TFOS DEWS II) in 2017, dry eye is a multifactorial disease of the ocular surface characterized by a loss of homeostasis of the tear film and is accompanied by ocular symptoms, in which tear film instability and hyperosmolarity, ocular surface inflammation and damage, and neurosensory abnormalities play etiological roles [40]. Based on these definitions, DED is a disease described by the disruption of homeostasis, whether in the tear film, anatomy, or the nervous system. Taken together, the main basis of DED is an insufficient amount of tears resulting in various uncomfortable problems on the ocular surface. In the DED model rats, our in vivo results demonstrated that the administration of PCE restored tear production and corneal irregularity, as well as increased the corneal goblet cell density. These findings indicate that PCE can improve eye health in DED by restoring the normal homeostasis of tears and improving the intrinsic properties of the tear film and the ocular surface.

5. Conclusions

In conclusion, PCE exhibited cytoprotective and anti-inflammatory effects in dry conditions induced by desiccation stress. Moreover, treatment with PCE in the animal model was effective at restoring the progression of DED by improving tear production, corneal irregularity, and goblet cell density. Our findings, thus, indicate that PCE can be used as a healthy functional food material for eye health, particularly in DED cases.

Author Contributions: J.-M.L., A.C., H.-H.L. and B.-H.K. carried out the experiments. B.-H.K. and S.J.P. devised and designed the experiments, analyzed the data, and drafted the manuscript. All authors have read and agreed to the published version of the manuscript.

Funding: This research was financially supported by the Ministry of SMEs and Startups (MSS), Korea, under the "Regional Specialized Industry Development Program R&D (S3270020)" supervised by the Korea Technology and Information Promotion Agency for SMEs (TIPA).

Institutional Review Board Statement: All animal studies were approved by the Institutional Animal Care and Use Committee (IACUC) of Chonbuk National University Hospital Non-Clinical Evaluation Center (IACUC approval No. 2022-42 dated on 17 February 2022).

Informed Consent Statement: Not applicable.

Data Availability Statement: The data are not publicly available due to privacy and ethical restrictions.

Conflicts of Interest: Authors were employed by the company MEDIENCE Co., Ltd. The authors declare that the research was conducted in the absence of any commercial or financial relationships that could be construed as a potential conflict of interest.

References

1. Craig, J.P.; Nichols, K.K.; Akpek, E.K.; Caffery, B.; Dua, H.S.; Joo, C.K.; Liu, Z.; Nelson, J.D.; Nichols, J.J.; Tsubota, K.; et al. TFOS DEWS II Definition and Classification Report. *Ocul. Surf.* **2017**, *15*, 276–283. [CrossRef] [PubMed]
2. Stapleton, F.; Alves, M.; Bunya, V.Y.; Jalbert, I.; Lekhanont, K.; Malet, F.; Na, K.S.; Schaumberg, D.; Uchino, M.; Vehof, J.; et al. TFOS DEWS II Epidemiology Report. *Ocul. Surf.* **2017**, *15*, 334–365. [CrossRef] [PubMed]
3. Galletti, J.G.; de Paiva, C.S. The ocular surface immune system through the eyes of aging. *Ocul. Surf.* **2021**, *20*, 139–162. [CrossRef] [PubMed]
4. Lemp, M.A.; Foulks, G.N. The definition and classification of dry eye disease: Report of the Definition and Classification Subcommittee of the International Dry Eye WorkShop. *Ocul. Surf.* **2007**, *5*, 75–92.
5. Davidson, H.J.; Kuonen, V.J. The tear film and ocular mucins. *Vet. Ophthalmol.* **2004**, *7*, 71–77. [CrossRef]
6. Zhang, X.; M, V.J.; Qu, Y.; He, X.; Ou, S.; Bu, J.; Jia, C.; Wang, J.; Wu, H.; Liu, Z.; et al. Dry Eye Management: Targeting the Ocular Surface Microenvironment. *Int. J. Mol. Sci.* **2017**, *18*, 1398. [CrossRef] [PubMed]
7. Sharma, A.; Hindman, H.B. Aging: A predisposition to dry eyes. *J. Ophthalmol.* **2014**, *2014*, 781683. [CrossRef]
8. de Paiva, C.S. Effects of Aging in Dry Eye. *Int. Ophthalmol. Clin.* **2017**, *57*, 47–64. [CrossRef]

9. Sheppard, J.D.; Nichols, K.K. Dry Eye Disease Associated with Meibomian Gland Dysfunction: Focus on Tear Film Characteristics and the Therapeutic Landscape. *Ophthalmol. Ther.* **2023**, *12*, 1397–1418. [CrossRef]
10. Javadi, M.A.; Feizi, S. Dry eye syndrome. *J. Ophthalmic. Vis. Res.* **2011**, *6*, 192–198.
11. Rao, S.K.; Mohan, R.; Gokhale, N.; Matalia, H.; Mehta, P. Inflammation and dry eye disease-where are we? *Int. J. Ophthalmol.* **2022**, *15*, 820–827. [CrossRef] [PubMed]
12. Zhuang, D.; Misra, S.L.; Mugisho, O.O.; Rupenthal, I.D.; Craig, J.P. NLRP3 Inflammasome as a Potential Therapeutic Target in Dry Eye Disease. *Int. J. Mol. Sci.* **2023**, *24*, 10866. [CrossRef] [PubMed]
13. Barabino, S.; Chen, Y.; Chauhan, S.; Dana, R. Ocular surface immunity: Homeostatic mechanisms and their disruption in dry eye disease. *Prog. Retin. Eye Res.* **2012**, *31*, 271–285. [CrossRef] [PubMed]
14. Stevenson, W.; Chauhan, S.K.; Dana, R. Dry eye disease: An immune-mediated ocular surface disorder. *Arch. Ophthalmol.* **2012**, *130*, 90–100. [CrossRef] [PubMed]
15. Lao, F.; Sigurdson, G.T.; Giusti, M.M. Health Benefits of Purple Corn (*Zea mays* L.) Phenolic Compounds. *Compr. Rev. Food Sci. Food Saf.* **2017**, *16*, 234–246. [CrossRef] [PubMed]
16. Cristianini, M.; Guillén Sánchez, J.S. Extraction of bioactive compounds from purple corn using emerging technologies: A review. *J. Food Sci.* **2020**, *85*, 862–869. [CrossRef]
17. Colombo, R.; Ferron, L.; Papetti, A. Colored Corn: An Up-Date on Metabolites Extraction, Health Implication, and Potential Use. *Molecules* **2021**, *26*, 199. [CrossRef]
18. Cai, T.; Ge-Zhang, S.; Song, M. Anthocyanins in metabolites of purple corn. *Front. Plant Sci.* **2023**, *14*, 1154535. [CrossRef]
19. Kim, H.J.; Kim, B.H.; Jin, B.R.; Park, S.J.; An, H.J. Purple Corn Extract Improves Benign Prostatic Hyperplasia by Regulating Prostate Cell Proliferation and Apoptosis. *Agric. Food Chem.* **2022**, *70*, 5561–5569. [CrossRef]
20. Barabino, S.; Labetoulle, M.; Rolando, M.; Messmer, E.M. Understanding Symptoms and Quality of Life in Patients With Dry Eye Syndrome. *Ocul. Surf.* **2016**, *14*, 365–376. [CrossRef] [PubMed]
21. Li, C.Y.; Kim, H.W.; Won, S.R.; Min, H.K.; Park, K.J.; Park, J.Y.; Ahn, M.S.; Rhee, H.I. Corn husk as a potential source of anthocyanins. *J. Agric. Food Chem.* **2008**, *56*, 11413–11416. [CrossRef]
22. Fernandez-Aulis, F.; Hernandez-Vazquez, L.; Aguilar-Osorio, G.; Arrieta-Baez, D.; Navarro-Ocana, A. Extraction and Identification of Anthocyanins in Corn Cob and Corn Husk from Cacahuacintle Maize. *J. Food Sci.* **2019**, *84*, 954–962. [CrossRef]
23. Kang, W.S.; Jung, E.; Kim, J. *Aucuba japonica* Extract and Aucubin Prevent Desiccating Stress-Induced Corneal Epithelial Cell Injury and Improve Tear Secretion in a Mouse Model of Dry Eye Disease. *Molecules* **2018**, *23*, 2599. [CrossRef] [PubMed]
24. Scarpellini, C.; Ramos Llorca, A.; Lanthier, C.; Klejborowska, G.; Augustyns, K. The Potential Role of Regulated Cell Death in Dry Eye Diseases and Ocular Surface Dysfunction. *Int. J. Mol. Sci.* **2023**, *24*, 731. [CrossRef] [PubMed]
25. Portal, C.; Gouyer, V.; Gottrand, F.; Desseyn, J.L. Ocular mucins in dry eye disease. *Exp. Eye Res.* **2019**, *186*, 107724. [CrossRef] [PubMed]
26. Baudouin, C.; Rolando, M.; Benitez Del Castillo, J.M.; Messmer, E.M.; Figueiredo, F.C.; Irkec, M.; Van Setten, G.; Labetoulle, M. Reconsidering the central role of mucins in dry eye and ocular surface diseases. *Prog. Retin. Eye Res.* **2019**, *71*, 68–87. [CrossRef] [PubMed]
27. Seo, Y.; Ji, Y.W.; Lee, S.M.; Shim, J.; Noh, H.; Yeo, A.; Park, C.; Park, M.S.; Chang, E.J.; Lee, H.K. Activation of HIF-1α (hypoxia inducible factor-1α) prevents dry eye-induced acinar cell death in the lacrimal gland. *Cell Death Dis.* **2014**, *5*, e1309. [CrossRef]
28. Barabino, S. Is dry eye disease the same in young and old patients? A narrative review of the literature. *BMC Ophthalmol.* **2022**, *22*, 85. [CrossRef] [PubMed]
29. Gaffney, E.A.; Tiffany, J.M.; Yokoi, N.; Bron, A.J. A mass and solute balance model for tear volume and osmolarity in the normal and the dry eye. *Prog. Retin. Eye Res.* **2010**, *29*, 59–78. [CrossRef]
30. Mohamed, H.B.; Abd El-Hamid, B.N.; Fathalla, D.; Fouad, E.A. Current trends in pharmaceutical treatment of dry eye disease: A review. *Eur. J. Pharm. Sci.* **2022**, *175*, 106206. [CrossRef]
31. Cevallos-Casals, B.A.; Cisneros-Zevallos, L. Stoichiometric and kinetic studies of phenolic antioxidants from Andean purple corn and red-fleshed sweetpotato. *J. Agric. Food Chem.* **2003**, *51*, 3313–3319. [CrossRef] [PubMed]
32. Wu, X.; Beecher, G.R.; Holden, J.M.; Haytowitz, D.B.; Gebhardt, S.E.; Prior, R.L. Concentrations of anthocyanins in common foods in the United States and estimation of normal consumption. *J. Agric. Food Chem.* **2006**, *54*, 4069–4075. [CrossRef] [PubMed]
33. Jing, P.; Giusti, M.M. Characterization of anthocyanin-rich waste from purple corncobs (*Zea mays* L.) and its application to color milk. *J. Agric. Food Chem.* **2005**, *53*, 8775–8781. [CrossRef] [PubMed]
34. Jing, P.; Giusti, M.M. Effects of extraction conditions on improving the yield and quality of an anthocyanin-rich purple corn (*Zea mays* L.) color extract. *J. Food Sci.* **2007**, *72*, C363–C368. [CrossRef] [PubMed]
35. Zhao, X.; Corrales, M.; Zhang, C.; Hu, X.; Ma, Y.; Tauscher, B. Composition and thermal stability of anthocyanins from chinese purple corn (*Zea mays* L.). *J. Agric. Food Chem.* **2008**, *56*, 10761–10766. [CrossRef]
36. Cuevas Montilla, E.; Hillebrand, S.; Antezana, A.; Winterhalter, P. Soluble and bound phenolic compounds in different Bolivian purple corn (*Zea mays* L.) cultivars. *J. Agric. Food Chem.* **2011**, *59*, 7068–7074. [CrossRef] [PubMed]
37. Zilić, S.; Serpen, A.; Akıllıoğlu, G.; Gökmen, V.; Vančetović, J. Phenolic compounds, carotenoids, anthocyanins, and antioxidant capacity of colored maize (*Zea mays* L.) kernels. *J. Agric. Food Chem.* **2012**, *60*, 1224–1231. [CrossRef]
38. Zhou, B.; Wang, X.; Guo, L.; Zhang, Z.; Xu, C. Toxicological assessment on safety of maize purple plant pigment. *J. Chin. Inst. Food Sci. Technol.* **2007**, *7*, 141–143.

39. Nabae, K.; Hayashi, S.M.; Kawabe, M.; Ichihara, T.; Hagiwara, A.; Tamano, S.; Tsushima, Y.; Uchida, K.; Koda, T.; Nakamura, M.; et al. A 90-day oral toxicity study of purple corn color, a natural food colorant, in F344 rats. *Food Chem. Toxicol.* **2008**, *46*, 774–780. [CrossRef]
40. Craig, J.P.; Nelson, J.D.; Azar, D.T.; Belmonte, C.; Bron, A.J.; Chauhan, S.K.; de Paiva, C.S.; Gomes, J.A.P.; Hammitt, K.M.; Jones, L.; et al. TFOS DEWS II Report Executive Summary. *Ocul. Surf.* **2017**, *15*, 802–812. [CrossRef]

nutrients MDPI

Article

Euscaphis japonica Kanitz Fruit Exerts Antiobesity Effects by Inhibiting the Early Stage of Adipogenic Differentiation

Eunbi Lee, Juhye Park and Ju-Ock Nam *

Department of Food Science and Biotechnology, Kyungpook National University 80, Daehakro, Bukgu, Daegu 702701, Republic of Korea; 21eunbi@knu.ac.kr (E.L.); pdw8606@knu.ac.kr (J.P.)
* Correspondence: namjo@knu.ac.kr; Tel.: +82-53-950-7760

Abstract: During the worldwide COVID-19 outbreak, there was an increase in the prevalence of obesity, including childhood obesity, due to which the awareness of obesity and interest in treatment increased. Accordingly, we describe EJF (*Euscaphis japonica* Kanitz fruit) extract as a candidate for naturally derived antiobesity agents. In this study, we found that EJF is involved in the early stage of adipogenic differentiation in vitro and finally inhibits adipogenesis. We propose two mechanisms for the antiobesity effect of EJF. First, EJF inhibits MDI-induced mitotic clonal expansion (MCE) by inducing cell cycle arrest at the initiation of adipogenic differentiation. The second aims to regulate stability and activation at the protein level of IRS1, which initiates differentiation in the early stage of differentiation. As a result, it was found that the activation of Akt decreased, leading to the inhibition of the expression of adipogenesis-related transcription factors (PPARγ, C/EBPα) and the subsequent suppression of adipogenic differentiation. In summary, we suggest that EJF can inhibit adipogenesis and lipid accumulation by suppressing the early stage of adipogenic differentiation in 3T3-L1 adipocytes. These findings indicate that EJF's functionality could be beneficial in the treatment of obesity, particularly childhood obesity associated with adipocyte hyperplasia.

Keywords: *Euscaphis japonica* Kanitz fruit; antiobesity; adipogenesis; 3T3-L1 preadipocytes; mitotic clonal expansion (MCE)

1. Introduction

During the worldwide COVID-19 outbreak, there was an increase in the prevalence of obesity due to increased food intake and decreased activity among children, adolescents, and young adults. In particular, the proportion of children and adolescents with obesity aged between 6 and 19 years significantly increased from 4% in 1975 to >18% in 2016, and the number of patients with childhood obesity has recently increased rapidly [1]. As the basic number of adipocytes is formed during childhood and adolescence, childhood obesity is accompanied by hyperplasia of adipocytes [2]. The complications of childhood obesity include hypertension, left ventricular hypertrophy, atherosclerosis, insulin resistance, diabetes, asthma, and nonalcoholic fatty liver, which are similar to those of adult obesity [3]. However, childhood obesity is considered a much greater risk factor than adult obesity. The period of growth and development in adolescence is associated with a normal increase in insulin resistance, but an additional insulin resistance due to obesity during this period causes glucose intolerance and may lead to type 2 diabetes [4]. Furthermore, because the enzymes that synthesize sex hormones are expressed in the adipose tissue, excessive adipose tissue can induce high androgen levels, causing hormonal imbalances. Consequently, girls with childhood obesity may experience menstrual abnormalities such as amenorrhea, irregular bleeding, and polycystic ovary syndrome, which can result in impaired fertility [5].

The currently available drugs for obesity treatment can be broadly classified into two categories, viz. medications that act on the central nervous system to influence appetite and medications that act on either the central nervous system or peripheral systems to promote thermogenesis, as well as drugs that act on the gastrointestinal tract to reduce absorption. However, these drugs can cause side effects, such as insomnia, dry mouth, tachycardia, constipation, and heart valve diseases [6,7]. Hence, there has been an increasing demand for new treatments due to the side effects of commercially available obesity drugs, resulting in the emergence of natural-based antiobesity agents [8,9]. In fact, natural-based antiobesity agents have gained popularity among consumers, particularly due to the general perception that they exert minimal or no side effects compared with pharmaceutical drugs. The commercial market for antiobesity agents derived from natural sources, which have been explored for their efficacy and safety, is of significant magnitude [10]. Therefore, we selected the fruit of *Euscaphis japonica* Kanitz as a candidate for natural-based antiobesity agents. Although the antiobesity effects of the leaf of *E. japonica Kanitz* have been reported, the molecular mechanisms related to adipogenic differentiation have not been completely elucidated [11], and the antiobesity effects of the fruit have not yet been reported. Therefore, in this study, we investigated whether *E. japonica* Kanitz fruit (EJF) extract exerts an inhibitory effect on adipocyte hyperplasia by inhibiting adipogenic differentiation for the treatment of obesity, especially childhood obesity. Through our study, we present EJF as a candidate antiobesity agent derived from natural products.

2. Materials and Methods

2.1. Preparation of Euscaphis japonica (Thunb.) Kanitz Fruit Extract

The plant extract (KPM018-040) used in this research was obtained from the Natural Product Central Bank at the Korea Research Institute of Bioscience and Biotechnology (Daejeon, Republic of Korea). The plant was collected from Seogwipo-si, Jeju-do, KOREA in 2002. A voucher specimen (KRIB 0000067) is kept in the herbarium of the Korea Research Institute of Bioscience and Biotechnology. The plant (50 g), dried in the shade and powdered, was added to 1 L of Methyl alcohol 99.9% (HPLC grade) and extracted through 30 cycles (40 KHz, 1500 W, 15 min. ultrasonication-120 min. standing per cycle) at room temperature using an ultrasonic extractor (SDN-900H, SD-ULTRASONIC CO., LTD, Seoul, Republic of Korea). After filtration (Qualitative Filter No.100, HYUNDAI MICRO CO., LTD, Seoul, Republic of Korea) and drying under reduced pressure, E. japonica extract (15.44 g) was obtained. For use in laboratory experiments, the obtained extract was dissolved in DMSO at a concentration of 50 mg/mL and further diluted before use.

2.2. Cell Culture

3T3-L1 preadipocytes were obtained from ATCC (Manassas, VA, USA) and were maintained in growth media consisting of DMEM-H (Gibco, Paisley, UK) with 25 mM glucose, 3.7 g/L NaHCO3, 1 mM sodium pyruvate, 10% bovine calf serum (*v*/*v*) (BCS; Gibco, Paisley, UK), and 1% penicillin–streptomycin (Gibco, Paisley, UK). The cells were incubated at 37 °C in a humidified atmosphere with 5% CO_2. When the cells reached 80% confluency in a 100 mm culture dish, they were seeded into a 6-well plate and allowed to grow for an additional 2 days to reach post-confluency before entering the differentiation stage. The 3T3-L1 preadipocytes used in this experiment were at passage numbers ranging from 7 to 13.

2.3. Adipogenic Differentiation of 3T3-L1 Preadipocytes

For the experiment, 3T3-L1 preadipocytes were seeded at a density of 1×10^5 cells/well. After 2–3 days, the cells reach 100% confluence. Following this, an additional 2 days are maintained to allow the cells to enter a post-confluence state, where they remain in a cell cycle arrest (G0/G1) state due to contact inhibition and can no longer proliferate. Following the post-confluence state, 3T3-L1 adipocytes were induced to differentiate. The medium was replaced with a medium containing MDI. The MDI used for adipogenic differentiation stands for the cocktail of chemical compounds composed of methylisobutylxanthine (IBMX), dexamethasone, and insulin. The medium consisted of DMEM-H (Gibco,

Paisley, UK) with 25 mM glucose, 3.7 g/L NaHCO3, 1 mM sodium pyruvate, 10% fetal bovine serum (*v*/*v*) (FBS; Gibco, Paisley, UK), 1% penicillin-streptomycin (Gibco, Paisley, UK), 1 μg/mL insulin, 0.25 mM dexamethasone, 0.5 mM IBMX, and 0.125 nM indomethacin (Sigma, USA, Saint Louis, MO, USA). After the second day of differentiation, the medium was switched to DMEM-H containing 10% FBS and only insulin to promote the maturation of adipocytes. We conducted our experiments using fully differentiated adipocytes on the 8th day of differentiation. When exploring the molecular levels associated with cell cycle arrest during the early stage of differentiation, they were treated with MDI and EJF for 16 and 24 h after 2 days post-confluence. They were then harvested and used in experiments. All media were replaced every 2 days.

2.4. Cell Viability Assay

The cytotoxicity of the extract was determined using Cell Counting Kit-8 (CCK-8; Dojindo Molecular Technologies, Kumamoto, Japan). 3T3-L1 cells were seeded into a 96-well plate at 1×10^4 cells/well and maintained for 24 h. Next, the cells were treated with the extract or DMSO for 48 h. Subsequently, CCK solution was added to each well and incubated for 1 h at 37 °C. The absorbance was measured at 450 nm using a microplate reader (Infinite F50; Tecan, Männedorf, Switzerland). For this experiment, EJF was treated at concentrations ranging from 1.58 μg/mL to 200 μg/mL. The group without EJF treatment was treated with 0.004% (*v*/*v*) DMSO.

2.5. Oil Red O Staining

According to the experimental design, 3T3-L1 preadipocytes were seeded into 6-well plates and treated with EJF to induce differentiation. On the 6th or 8th day of differentiation, the adipocytes were washed with PBS. Then, they were fixed by treating with 1 mL of 4% paraformaldehyde (Biosesang Inc., Gyeonggi-do, Republic of Korea) for 1 h at room temperature. Subsequently, they were stained with 0.6% Oil Red O solution (Sigma, USA, Saint Louis, MO, USA) at room temperature for 30 min. After staining, the product is washed three times with third-distilled water, and then 1 mL of isopropyl alcohol (Duksan Pure Chemicals, Korea) is added to a 6-well plate. The plate is shaken at 80 rpm for 10 min at room temperature to dissolve the final product (ORO content). Subsequently, the absorbance is measured at 450 nm using a microplate reader in 200 μL aliquots on a 96-well plate.

2.6. Real-Time Reverse Transcription Polymerase Chain Reaction (RT-PCR)

3T3-L1 cells were seeded into 6-well plates and treated with EJF according to the experimental design to induce differentiation, after which they were washed once with PBS. Total RNA was isolated using Trizol reagent, and then cDNA libraries were synthesized using the PrimeScript^{TM} RT Reagent Kit (TaKaRa Bio, Kyoto, Japan). For the synthesis of the cDNA library, standard PCR was performed with the following settings: pre-incubation (15 min at 37 °C), annealing (5 min at 50 °C), extension (5 min at 98 °C), and cooling (4 °C). The mRNA expression levels were measured and analyzed using the iCycleriQTM Real-Time PCR Detection System (Bio-Rad Laboratories, Hercules, CA, USA), with detection performed using SYBR Green (TOYOBO, Japan). The specific thermal cycling conditions used for RT-PCR were as follows: pre-incubation (1 min at 95 °C), amplification (15 s at 95 °C, followed by 1 min at 60 °C for 39 cycles), melting (10 s at 95 °C), and cooling (5 s at 72.5 °C). The mRNA expression levels were normalized to those of β-actin and indicated as fold changes compared with a control group. All experiments were performed biologically and technically in triplicate. The primers used for RT-PCR were synthesized by Macrogen (Seoul, Republic of Korea), and their sequences are listed in Table 1.

Table 1. Primer sequences and corresponding accession numbers used for qPCR.

Gene Name	Accession No.		Sequence
Adipoq	NM_009605	Forward	5′-ACCTACGACCAGTATCAGGAAAAG-3′
		Reverse	3′-ACTAAGCTGAAAGTGTGTCGACTG-5′
C/ebpα	NM_001287523	Forward	5′-TTACAACAGGCCAGGTTTCC-3′
		Reverse	3′-GGCTGGCGACATACAGATCA-5′
Pparγ	AB644275	Forward	5′-TTTTCAAGGGTGCCAGTTTC-3′
		Reverse	3′-AATCCTTGGCCCTCTGAGAT-5′
β-actin	EF095208	Forward	5′-GACAACGGCTCCGGCATGTGCAAAG-3′
		Reverse	3′-TTCACGGTTGGCCTTAGGGTTCAG-5′

2.7. Western Blot Assay

3T3-L1 cells were lysed in RIPA buffer supplemented with 1× phosphatase cocktail and protease cocktail. The lysate was then mixed with SDS to produce a protein sample, which was boiled at 100 °C for 10 min. Protein quantification was performed using the Bradford method. A 30 µg total protein sample was loaded onto 7.5–15% SDS-polyacrylamide gel for the purpose of separation. Subsequently, the proteins were transferred to nitrocellulose membranes, which were then blocked with a solution of 5% nonfat skim milk in TBS-T buffer (10 mM Tris pH 8.0, 150 mM NaCl, and 0.05% Tween 20) for 1 h at room temperature with shaking at 80 rpm. Primary antibodies were applied and incubated overnight at 4 °C. After three washes with TBS-T buffer, the membranes were incubated with horseradish peroxidase-conjugated secondary antibody for 1 h at room temperature. The protein bands with bound antibodies were detected using an enhanced chemiluminescence kit (GE Healthcare, Buckinghamshire, UK) and visualized using the Fusion Solo Detector. The β-actin and β-tubulin bands were used as a normalization control for the targeted protein bands. The antibodies used for protein detection are listed in Table 2.

Table 2. Antibody information used for Western blotting.

Gene Name	Company	Product No.	IgG	Conc.
		Primary antibody		
PPARγ	SCBT	sc-7273	M	1:500
C/EBPα	CST	2295S	R	1:1000
IRS1	CST	2390S	R	1:1000
p-IRS1	Invitrogen	2103503	R	1:1000
Akt	CST	4691S	R	1:1000
p-Akt	CST	4060S	R	1:1000
β-actin	SCBT	sc-47778	M	1:500
β-tubulin	Abcam	ab179513	R	1:1000
		Secondary antibody		
Mouse	CST	7076S		1:2000
Rabbit	CST	7074S		1:3000

2.8. Cell Cycle Assay

To evaluate the cell cycle, 3T3-L1 adipocytes treated with MDI and the extract were separated into single cells using Trypsin^{TM} Express Enzyme (TE; Thermofisher Scientific, Waltham, MA, USA), which consists of Trypsin (1X), phenol Red, and EDTA. To examine the cell cycle distribution based on the difference in fluorescence expression levels within a single adipocyte, we detached the adipocytes attached to the well by treating them with TE. Subsequently, we generated single cells through multiple pipetting steps. Single cells were then fixed with cold 95% ethanol and harvested by centrifugation at 3000× *g* for 5 min. The resulting cell pellet was resuspended in a solution containing propidium iodide (100 µg/mL) and Rnase A (100 µg/mL) for staining. Stained cells were incubated at 37 °C for 30 min in a CO_2 incubator. Cell cycle analysis was performed using an Attune acoustic focusing cytometer (Thermofisher Scientific, Waltham, MA, USA). For the experiment, we set the voltage to 80 for Forward Scatter (FSC) and 260 for Side Scatter (SSC). After gating, we counted 1×10^4 cells and measured them using a 488 nm excitation laser and a 574/26 emission filter (nm).

2.9. Statistical Analysis

The experimental results were analyzed using the GraphPad Prism 9.4.1 software (Sandiego, CA, USA) for data processing. All experiments were independently repeated at least three times, and the data are presented as the mean ± standard deviation (SD) of individual measurements within independent experiments. Statistical analysis was conducted using IBM SPSS statistics 25. One-way analysis of variance followed by Tukey's post hoc comparison test was performed, and statistical significance was set at $p < 0.05$.

3. Results

3.1. EJF Extract Inhibits Adipogenic Differentiation in 3T3-L1 Preadipocytes

Through this section, we confirmed that EJF inhibits adipogenic differentiation and lipid accumulation in 3T3-L1 adipocytes. Before initiating differentiation, we determined the optimal treatment dose of EJF through cell viability. Since no significant cytotoxicity was observed up to a dose of 50 µg/mL, it was chosen as the highest dose (Figure 1A). Subsequently, adipogenic differentiation was induced over an 8-day period with or without EJF treatment. The scheme used for differentiation is depicted in Figure 1B. As a result, EJF treatment significantly decreased both adipogenic differentiation and lipid accumulation. Furthermore, these effects were found to be dose-dependent. As adipogenic differentiation is regulated by the transcriptional cascade involving factors such as PPARγ and C/EBPα [1], we examined the effect of EJF extract on the expression of adipogenesis-related transcription factors. Adiponectin, secreted by mature adipocytes, can be used as a marker for in vitro differentiation. Therefore, we also investigated the mRNA expression of adiponectin. The protein levels of PPARγ and C/EBPα showed minimal changes at 12.5 and 25 µg/mL EJF concentrations but significantly decreased at 50 µg/mL (Figure 1D,E). Similarly, at the mRNA level, the expression levels of Pparγ, C/ebpα, and Adipoq decreased in a dose-dependent manner (Figure 1F). These findings demonstrate that EJF inhibits the differentiation of preadipocytes into mature adipocytes and impedes lipid accumulation in mature adipocytes.

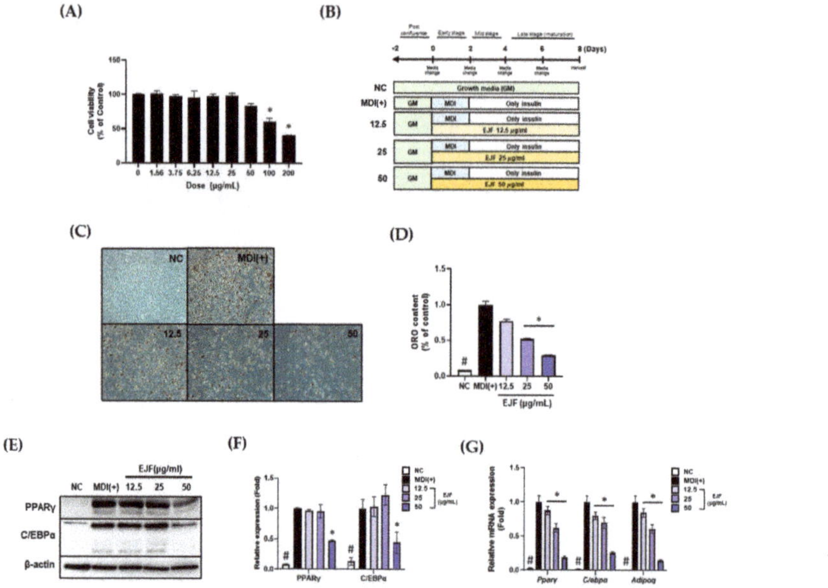

Figure 1. Effects of *Euscaphis japonica* Kanitz fruit (EJF) extract on fully differentiated 3T3-L1 adipocytes.

(**A**) A CCK assay was conducted to select a noncytotoxic concentration of EJF before treating 3T3-L1 preadipocytes. Cell viability was considered noncytotoxic if it was >80% of that in the control group, and that concentration was used in the experiments. (**B**) Scheme of adipogenic differentiation for this experiment. (**C**) 3T3-L1 preadipocytes were treated with EJF during the differentiation period. Representative images of each group were obtained through Oil Red O (ORO) staining to evaluate the extent of lipid accumulation. (**D**) Lipid droplets stained by Oil Red O (ORO) staining were quantified by dissolving them in isopropanol and measuring the absorbance at 450 nm. (**E**) Protein expression levels of PPARγ and C/EBPα in fully differentiated 3T3-L1 adipocytes were detected by Western blotting. (**F**) The Western blot bands representing the protein expression levels of PPARγ and C/EBPα were quantified and presented graphically. β-actin values were used to confirm equal loading, and the expression levels of all factors were normalized to those of β-actin. (**G**) mRNA expression levels of Pparγ, C/ebpα, and Adipoq in fully differentiated 3T3-L1 adipocytes were detected using RT-PCR. Equal loading was confirmed by β-actin, and the expression levels of all factors were normalized to β-actin values. All experiments were independently repeated three times, and data are expressed as mean ± SD. * $p < 0.05$ compared with the MDI (+) group. # $p < 0.01$ when compared with the NC group.

3.2. The Inhibitory Effect of EJF on Adipogenesis Is Most Critical in the Early Stage of Differentiation in 3T3-L1 Preadipocytes

Through this section, we discovered that the inhibitory effect of EJF on adipogenesis is most critical during the early stage of differentiation. EJF inhibits adipogenic differentiation and lipid accumulation, leading to the investigation of the stage at which EJF intervenes. The differentiation of adipocytes can be divided into early stage, middle stage, terminal differentiation depending on the type of cocktail and duration of differentiation. Based on the information below, we divided the differentiation stages into three phases [12]

- Early stage: After post-confluence, treatment of MDI initiates differentiation. This stage is characterized by the presence of growth-arrested preadipocytes and mitotic clonal expansion (MCE) phase.
- Middle stage: After 2 days of MDI treatment, transitioning to media containing only insulin leads to entering the middle stage. In this stage, adipogenic gene expression is initiated, and the accumulation of lipid droplets begins.
- Late stage: After the 4th of differentiation, adipocytes enter the maturation stage. This stage is characterized by the formation of mature lipid-filled adipocytes.

Firstly, we have designed a scheme to treat EJF at different stages of differentiation (Figure 2A). EJF was administered at a dose of 50 μg/mL. The results revealed a significant difference in the level of differentiation between Group A, which did not undergo EJF treatment, and Group B, which received six days of treatment. Specifically, among the groups treated with EJF for two days (Groups C, E, and G), the differentiation decrease in Group C was approximately 50% lower in terms of ORO content compared to the control group (Group A). However, no significant differentiation decrease was observed in Groups E and G, which received EJF treatment during the middle and late stages, compared to Group C. Furthermore, among the groups treated with EJF for four days (Groups D, F, and H), there was a significant difference in the differentiation between the groups (D and H) that underwent the early stage and the group F that did not undergo the early stage (Figure 2B,C). These results indicate that EJF critically regulates the early stage of differentiation regardless of the treatment duration.

Additionally, the expression level of Pparγ, a master transcription factor that induces adipogenic differentiation, was investigated. When comparing groups with the same total extract treatment duration, similar to the method used for comparing ORO content, the expression of Pparγ was significantly decreased in the groups treated with EJF during the early stage compared to the groups without early-stage treatment (Figure 2D). This is consistent with the trend observed in ORO content. Therefore, our findings suggest that

EJF strongly regulates the early stage of adipogenic differentiation, leading to a decrease in lipid accumulation.

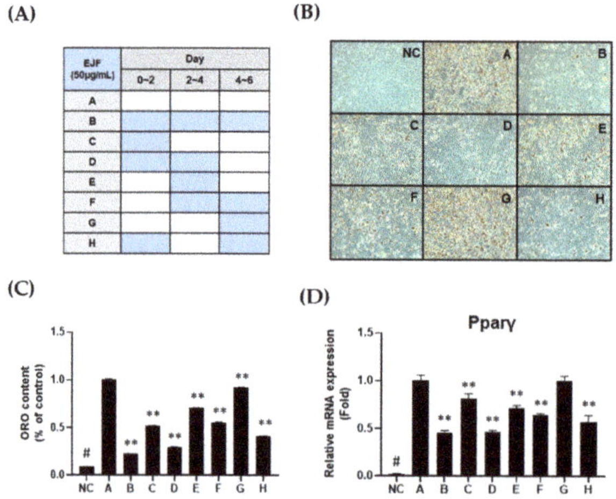

Figure 2. Effects of *Euscaphis japonica* Kanitz fruit (EJF) extract on each stage of adipogenic differentiation. (**A**) Scheme for EJF treatment during the differentiation period of 3T3-L1 preadipocytes. 3T3-L1 preadipocytes were differentiated for 6 days in the differentiation medium containing MDI, depending on the presence or absence of EJF. (**B**) Mature adipocytes differentiated for 6 days were stained for intracellular lipid droplets using Oil Red O (ORO) staining. (**C**) The stained adipocytes were quantified by measuring the absorbance at 450 nm after dissolving them in isopropanol. (**D**) Total RNA was extracted from the 6-day-differentiated adipocytes to measure Pparγ expression using real-time PCR. Equal loading was confirmed using β-actin, and the expression levels of all factors were normalized to those of β-actin. All experiments were performed independently in triplicate, and data are expressed as mean ± SD. ** $p < 0.01$ compared with the A group. # $p < 0.01$ when compared with the NC group.

3.3. EJF Inhibits the Early Stage of Adipogenic Differentiation by Inhibiting MDI-Induced Cell Cycle Progression in 3T3-L1 Preadipocytes

In this section, we discovered that EJF delays cell cycle progression by inducing MDI during the early stage of differentiation. For the progression of differentiation, 3T3-L1 preadipocytes stop their growth through contact inhibition due to inhibitory signals from nearby cells. In preadipocytes with growth cessation, the level of p27 KIP1 protein increases, and the expression levels of cyclin-dependent kinase 4 (CDK4) and cyclin D decrease, resulting in cell cycle arrest in the G0/G1 phase. When MDI is administered in the cell cycle arrest state, MCE (S phase) is initiated after approximately 14 h. Through MCE, the cell cycle of adipocyte is resumed, and adipocytes prepare for lipid formation and accumulation in subsequent stages. Therefore, delaying the initiation of MCE by blocking cell cycle progression may be an efficient method to inhibit adipogenesis [2]. EJF plays a key role in adipogenic differentiation in the early stage, leading to the investigation of its effect on MCE initiation. Based on the MCE initiation point, we have specified certain time points as 16 and 24 h. We conducted the experiment by dividing the groups based on the presence or absence of MDI and EJF as follows: growth-arrested preadipocytes (NC), MDI (+), and EJF-treat groups categorized by concentration (μg/mL). Afterward, we examined the changes in cell cycle distribution and the expression of cell cycle-related genes after EJF treatment, and the results are as follows.

- Cell cycle distribution: The NC group induced cell cycle arrest in G0/G1 phase, and the MDI (+) group initiated both cell cycle progression and MCE (S phase). However, after EJF treatment, the percentage of cells in S phase decreased and the percentage of cells in G0/G1 increased (Figure 3A,B).

- CDK4, Cyclin D: A significant decrease in the expression levels of CDK4 and cyclin D was observed at 16 h in the group treated with EJF compared to that in the group without EJF treatment (Figure 3C–F).
- p27 KIP1: The MDI (+) group exhibited a decrease in the expression of p27KIP1 compared to the NC group due to the resumption of the cell cycle. However, treatment with EJF significantly increased the expression of p27KIP1, and this observation remained consistent regardless of the time point (Figure 3C,D,G).

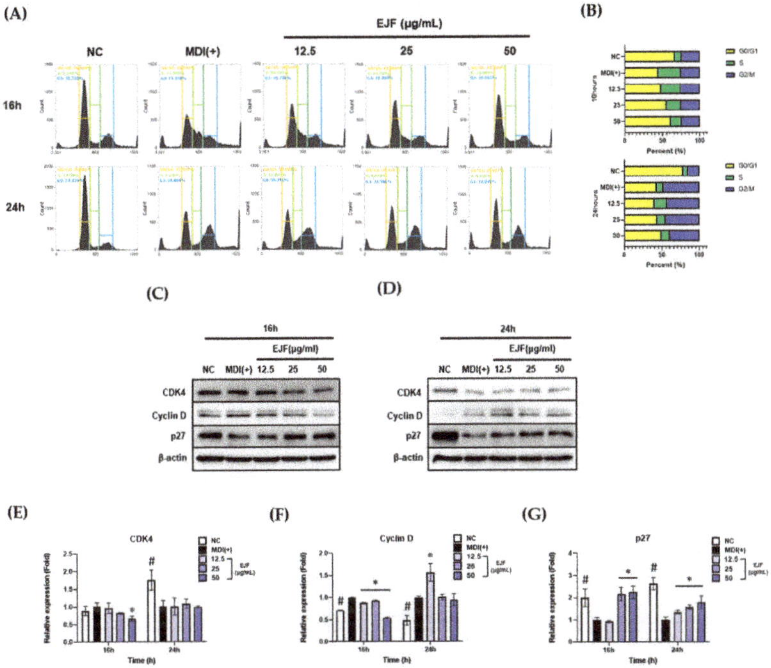

Figure 3. Effects of *Euscaphis japonica* Kanitz fruit (EJF) extract on MDI-induced cell cycle progression in 3T3-L1 preadipocytes. 3T3-L1 preadipocytes were treated with MDI-containing medium for 16 and 24 h, following cell cycle arrest induced by contact inhibition, depending on the presence or absence of EJF. (**A**) 3T3-L1 cells were fixed with 95% ethanol and then treated with propidium iodide. The cell population was determined using a flow cytometer, with a count of 1×10^4 cells per sample. (**B**) The distribution of the quantified cell population was represented graphically. (**C**) After treating the cells with EJF in MDI-containing medium for 16 h, the expression levels of CDK4, cyclin D, and p27 KIP1 proteins were measured by Western blotting. (**D**) After treating the cells with EJF in MDI-containing medium for 24 h, the expression levels of CDK4, cyclin D, and p27 KIP1 proteins were examined by Western blotting. Equal loading of proteins was confirmed by the expression of β-actin. (**E–G**) The Western blot bands obtained from cells treated with EJF or control in MDI-containing medium for 16 and 24 h were quantified and presented graphically. The expression levels of all factors were normalized to those of β-actin. All experiments were independently performed in triplicate, and data are expressed as mean ± SD. * $p < 0.05$ compared with the MDI (+) group. # $p < 0.01$ when compared with the NC group.

This trend was more significant at 16 h than at 24 h. Therefore, this result suggests that EJF treatment delays MDI-induced cell cycle resumption—that is, entry into MCE.

3.4. *EJF Attenuates IRS1 Stability and Inhibits Its Activation in 3T3-L1 Preadipocytes*

In this section, we discover that EJF reduces the stability and activation of IRS1 and subsequently reduces the activation of Akt. When treating adipocytes with MDI for adipogenic differentiation, insulin first binds to the insulin receptor on adipocytes. Subsequently, insulin receptor substrate 1 (IRS1) becomes phosphorylated and activated by the receptor. Activated IRS1 then converts phosphoinositide 3-kinase (PI3K) into phosphatidylinositol 4,5-bisphosphate and phosphatidylinositol 3,4,5-trisphosphate, leading to the phosphorylation of Akt. Akt is then recruited to the cell membrane and becomes activated. Therefore, this signaling cascade directly participates in cellular differentiation and adipogenesis by regulating adipogenesis-related genes. In this process, IRS1, which directly modulates the response to external MDI and subsequent differentiation, exerts a significant influence on the initiation of the early stages of differentiation [11]. EJF plays a key role in adipogenic differentiation in the early stage, leading to the investigation of its effect on IRS1 phosphorylation and activation. We established the hypothesis that EJF treatment would decrease IRS1 phosphorylation and stability, subsequently inhibiting Akt phosphorylation, ultimately leading to the suppression of adipogenesis when inducing differentiation using MDI-containing medium. Phosphorylation of IRS1 was significantly increased in the MDI (+) group compared to the NC group. However, EJF treatment decreased both IRS1 stability and IRS1 phosphorylation in a dose-dependent manner. These results suggest that EJF treatment significantly reduces IRS1 phosphorylation and activation. Next, phosphorylation of Akt was also significantly inhibited in the 50 µg/mL group compared to the MDI (+) group (Figure 4A,B). In summary, EJF inhibits adipogenic differentiation with a strategy to decrease phosphorylation and the stability of IRS1 and inhibit Akt phosphorylation in the early stage of differentiation.

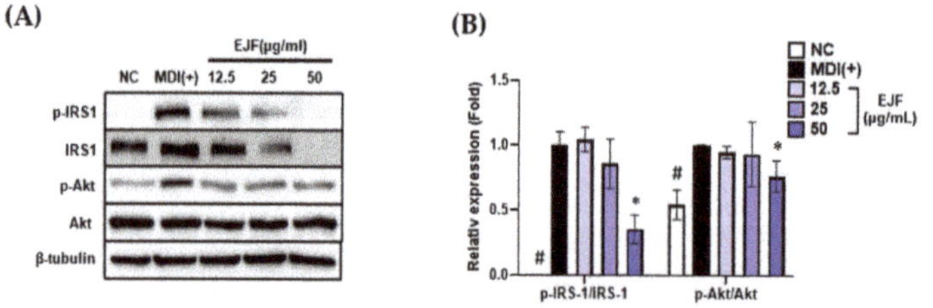

Figure 4. Effects of *Euscaphis japonica* Kanitz fruit (EJF) extract on IRS1 stability and its activation during the early stage of adipogenic differentiation. 3T3-L1 preadipocytes were treated with medium containing MDI for 30 min, depending on the presence or absence of EJF. (**A**) After treating 3T3-L1 cells for 30 min, the expression levels of p-IRS1, IRS1, p-Akt, and Akt proteins were detected by Western blotting. (**B**) The levels of phosphorylated IRS1 and its downstream target Akt phosphorylation were quantified by analyzing Western blot bands and presented graphically. Equal loading was confirmed by β-tubulin, and the expression levels of all factors were normalized to β-tubulin values. All experiments were independently repeated three times, and data were expressed as mean ± SD. * $p < 0.05$ compared with the MDI (+) group. # $p < 0.01$ when compared with the NC group.

4. Discussion

In this study, two mechanisms were proposed to explain the antiobesity effects of EJF during the early stage of adipogenic differentiation. The first mechanism involves the delay in differentiation progression through the inhibition of MCE via cell cycle arrest. The second mechanism involves a decrease in IRS1 stability and reduced phosphorylation of IRS1. Adipogenic differentiation is a highly regulated process involving the regulation of gene expression, induction of transcription factors that drive adipocyte development, and cascades of cell cycle proteins [12]. The MDI-induced differentiation model of murine 3T3-L1 preadipocytes is widely used in obesity research [13]. The early stage of adipogenic differentiation can be largely divided into growth arrest through contact inhibition, MCE, and postmitotic growth arrest stages [14]. When preadipocytes proliferate and reach a high level of confluency, growth contact inhibition occurs within the cells, wherein no more cell growth occurs, and cell cycle arrest continues. In this state, when stimulated by hormones such as insulin and dexamethasone, the cells undergo reentry into the cell cycle, resulting in MCE. After a certain period of time following MCE, another round of growth arrest occurs, resulting in the subsequent attainment of terminal differentiation [15]. During cell cycle arrest and progression, preadipocytes overexpress p27 KIP1 in the first cell cycle arrest and remain in the G1 phase. During the early stages of differentiation, cyclin D is activated and assembled by CDK4. The cyclin D–CDK4 complex acts as a regulator in the early G1 phase of the cell cycle, facilitating the progression of cells from the G1 phase to the S phase [16]. Therefore, inhibiting the progression to MCE by delaying the cell cycle progression of preadipocytes through the regulation of cell cycle-related proteins can be a novel strategy to inhibit adipogenic differentiation. Moreover, when insulin stimulation for the induction of differentiation is provided, insulin binds to the alpha and beta subunits of the insulin receptor in the adipocyte membrane and induces phosphorylation [17], followed by tyrosine phosphorylation of IRS1 [18]. Phosphorylated IRS1 induces the activation of MEK, ERK, and Akt, and these factors regulate adipogenic differentiation [19]. IRS1, which plays such an important role in adipogenic differentiation, is degraded by E3 ligase activated by stimuli such as extract treatment [20]. Hence, reducing the stability or phosphorylation of IRS1 in the early stage of differentiation can be a strategy to suppress adipogenesis.

After being regulated by these upstream factors, adipogenic differentiation is finally regulated by adipogenesis-related transcription factors. PPARγ is a transcription factor that strongly induces adipogenesis by inducing lipid accumulation, morphological changes in adipocytes, adipose tissue-specific gene expression, and cell growth arrest [21]. C/EBPα is one type of CCAAT/enhancer-binding protein and is widely recognized as a transcription factor involved in adipogenic differentiation. Similar to PPARγ, C/EBPα also acts as a master regulator of adipogenic differentiation, maintaining the phenotype of mature adipocytes and cooperatively inducing the expression of adipose tissue-specific genes along with PPARγ [22]. Therefore, we selected PPARγ and C/EBPα as markers for adipogenic differentiation and adiponectin as a marker representing the extent of final adipogenic differentiation. Adiponectin is a hormone secreted by mature adipocytes, and it plays a role in glucose regulation and fatty acid oxidation within the body [23]. However, in the present study, adiponectin was used as a marker to evaluate the degree of adipogenic differentiation and lipid accumulation. EJF treatment regulated the early stage of adipogenic differentiation through these strategies and ultimately reduced the number of mature adipocytes by inhibiting the expression of adipogenesis-related transcription factors throughout the entire process of adipocyte development. In the case of childhood obesity, which is often accompanied by adipocyte hyperplasia, these mechanisms of action of EJF are particularly advantageous for treatment [2].

We suggest further research to supplement our study findings. It is important to identify the bioactive components present in EJF that contribute to the effects observed in this study. Previous studies have reported the presence of euscaphinin and euscapholide in the leaves [24], triterpene acid in the branches [25], and hexacyclic triterpenic acid in the roots [26] of *E. japonica* Kanitz. However, there have been no reports on the components present in the fruit of this plant. Therefore, we intend to analyze and quantify the components present in EJF using MS BPI chromatogram and UPLC-QTOF-MS. Subsequently, we shall conduct in vitro experiments using the fractions obtained from the chromatogram to select one fraction that exhibits similar adipogenic inhibitory activities to those of the original extract. We shall then identify the main component within that fraction using nuclear magnetic resonance [27].

5. Conclusions

We discovered that EJF inhibits adipogenesis by inhibiting the early stage of adipogenic differentiation, including the initiation of differentiation. In this regard, we have proposed two major mechanisms for inhibiting adipogenic differentiation induced by MDI. The first mechanism is the delay in entry into MCE due to the induction of cell cycle arrest during the early stage of differentiation. This was demonstrated by cell cycle arrest induced by increased expression of p27KIP1 and decreased expression of CDK4 and Cyclin D compared to the MDI (+) group, resulting in increased G0/G1 distribution. The second is the decreased stability and activation of IRS1 during the initiation stage of differentiation. This was demonstrated by the reduction of IRS1 expression and phosphorylation due to EJF treatment when IRS1 activation was induced by MDI. This resulted in decreased phosphorylation of Akt, leading to the suppression of the expression of PPARγ and C/EBPα, which are master regulators of adipogenic differentiation [28]. As a result, the delay in entry into MCE and decreased stability and activation of IRS1 ultimately reduced the extent of adipogenesis and lipid accumulation during subsequent stages of differentiation. Therefore, through this study, we have demonstrated that EJF effectively inhibits excessive adipocyte formation in the early stage of differentiation, suggesting EJF as a potential candidate for the prevention and treatment of childhood obesity, which is accompanied by adipocyte hyperplasia. Our study can contribute to identifying bioactive components with these effects in EJF to develop a treatment for childhood obesity; thus, it can contribute to treating childhood obesity, which has increased due to the COVID-19 pandemic.

Author Contributions: E.L. and J.P. performed experiments; J.-O.N. supervised the project; E.L., J.P. and J.-O.N. wrote the manuscript. All authors have read and agreed to the published version of the manuscript.

Funding: This paper did not receive any funding support.

Institutional Review Board Statement: Not applicable.

Informed Consent Statement: Not applicable.

Data Availability Statement: The data presented in this study are available on request from the corresponding author.

Conflicts of Interest: The authors declare no conflict of interest.

References

1. Stavridou, A.; Kapsali, E.; Panagouli, E.; Thirios, A.; Polychronis, K.; Bacopoulou, F.; Psaltopoulou, T.; Tsolia, M.; Sergentanis, T.N.; Tsitsika, A. Obesity in children and adolescents during COVID-19 pandemic. *Children* **2021**, *8*, 135. [CrossRef] [PubMed]
2. Suwa, A.; Kurama, T.; Shimokawa, T. Adipocyte hyperplasia and RMI1 in the treatment of obesity. *FEBS J.* **2011**, *278*, 565–569. [CrossRef] [PubMed]
3. Freemark, M. Childhood obesity in the modern age: Global trends, determinants, complications, and costs. In *Pediatric Obesity*; Springer: Berlin/Heidelberg, Germany, 2018; pp. 3–24.
4. Daniels, S. Complications of obesity in children and adolescents. *Int. J. Obes.* **2009**, *33*, S60–S65. [CrossRef]
5. Robinson, G.A.; Geier, M.; Rizzolo, D.; Sedrak, M. Childhood obesity: Complications, prevention strategies, treatment. *J. Am. Acad. Physician Assist.* **2011**, *24*, 58–65. [CrossRef] [PubMed]
6. Yoo, S.J. Pharmacological Treatment of Obesity. In *Endocrinology and Diabetes*; Springer: Berlin/Heidelberg, Germany, 2008; p. 23.
7. Ioannides-Demos, L.L.; Proietto, J.; Tonkin, A.M.; McNeil, J.J. Safety of drug therapies used for weight loss and treatment of obesity. *Drug Saf.* **2006**, *29*, 277–302. [CrossRef] [PubMed]
8. Vasudeva, N.; Yadav, N.; Sharma, S.K. Natural products: A safest approach for obesity. *Chin. J. Integr. Med.* **2012**, *18*, 473–480. [CrossRef]
9. Fu, C.; Jiang, Y.; Guo, J.; Su, Z. Natural products with anti-obesity effects and different mechanisms of action. *J. Agric. Food Chem.* **2016**, *64*, 9571–9585. [CrossRef]
10. Vermaak, I.; Viljoen, A.M.; Hamman, J.H. Natural products in anti-obesity therapy. *Nat. Prod. Rep.* **2011**, *28*, 1493–1533. [CrossRef]
11. Kim, K.H.; Choi, S.H.; Lee, T.S.; Oh, W.K.; Kim, D.S.; Kim, J.B. Selective LXRα inhibitory effects observed in plant extracts of MEH184 (*Parthenocissua tricuspidata*) and MEH185 (*Euscaphis japonica*). *Biochem. Biophys. Res. Commun.* **2006**, *349*, 513–518. [CrossRef]
12. Moreno-Navarrete, J.M.; Fernández-Real, J.M. Adipocyte differentiation. In *Adipose Tissue Biology*; Springer: Berlin/Heidelberg, Germany, 2017; pp. 69–90.
13. Knutson, V.P. 3T3-L1 adipocytes as a cell culture model of insulin resistance. *In Vitro Cell. Dev. Biol. Anim.* **1997**, *33*, 77–81. [CrossRef]

14. Chang, E.; Kim, C.Y. Natural products and obesity: A focus on the regulation of mitotic clonal expansion during adipogenesis. *Molecules* **2019**, *24*, 1157. [CrossRef]
15. Fajas, L. Adipogenesis: A cross-talk between cell proliferation and cell differentiation. *Ann. Med.* **2003**, *35*, 79–85. [CrossRef]
16. Park, H.; Cho, J.A.; Lim, E.H.; Lee, C.W.; Lee, S.H.; Seo, S.W.; Yang, D.Y.; Lee, K.W. Cell cycle regulators are critical for maintaining the differentiation potential and immaturity in adipogenesis of adipose-derived stem cells. *Differentiation* **2011**, *82*, 136–143. [CrossRef]
17. Meng, Q.; Qi, X.; Chao, Y.; Chen, Q.; Cheng, P.; Yu, X.; Kuai, M.; Wu, J.; Li, W.; Zhang, Q. IRS1/PI3K/AKT pathway signal involved in the regulation of glycolipid metabolic abnormalities by Mulberry (*Morus alba* L.) leaf extracts in 3T3-L1 adipocytes. *Chin. Med.* **2020**, *15*, 1. [CrossRef]
18. Liu, T.; Yu, B.; Kakino, M.; Fujimoto, H.; Ando, Y.; Hakuno, F.; Takahashi, S.-I. A novel IRS-1-associated protein, DGKζ regulates GLUT4 translocation in 3T3-L1 adipocytes. *Sci. Rep.* **2016**, *6*, 35438. [CrossRef]
19. Maki, C.; Funakoshi-Tago, M.; Aoyagi, R.; Ueda, F.; Kimura, M.; Kobata, K.; Tago, K.; Tamura, H. Coffee extract inhibits adipogenesis in 3T3-L1 preadipocyes by interrupting insulin signaling through the downregulation of IRS1. *PLoS ONE* **2017**, *12*, e0173264. [CrossRef] [PubMed]
20. Yoneyama, Y.; Inamitsu, T.; Chida, K.; Iemura, S.-I.; Natsume, T.; Maeda, T.; Hakuno, F.; Takahashi, S.-I. Serine phosphorylation by mTORC1 promotes IRS-1 degradation through SCFβ-TRCP E3 ubiquitin ligase. *iScience* **2018**, *5*, 1–18. [CrossRef] [PubMed]
21. Spiegelman, B.; Puigserver, P.; Wu, Z. Regulation of adipogenesis and energy balance by PPARγ and PGC-1. *Int. J. Obes.* **2000**, *24*, S8–S10. [CrossRef] [PubMed]
22. Lee, J.-E.; Schmidt, H.; Lai, B.; Ge, K. Transcriptional and epigenomic regulation of adipogenesis. *Mol. Cell. Biol.* **2019**, *39*, e00601-18. [CrossRef]
23. Ghoshal, K.; Chatterjee, T.; Chowdhury, S.; Sengupta, S.; Bhattacharyya, M. Adiponectin genetic variant and expression coupled with lipid peroxidation reveal new signatures in diabetic dyslipidemia. *Biochem. Genet.* **2021**, *59*, 781–798. [CrossRef]
24. Maeda, H.; Matsuo, Y.; Tanaka, T.; Kouno, I. Euscaphinin, a New Ellagitannin Dimer from Euscaphis japonica (T HUNB.) K ANITZ. *Chem. Pharm. Bull.* **2009**, *57*, 421–423. [CrossRef] [PubMed]
25. Zhang, L.-J.; Cheng, J.-J.; Liao, C.-C.; Cheng, H.-L.; Huang, H.-T.; Kuo, L.-M.Y.; Kuo, Y.-H. Triterpene acids from Euscaphis japonica and assessment of their cytotoxic and anti-NO activities. *Planta Med.* **2012**, *78*, 1584–1590. [CrossRef]
26. Chen, Y.; Zhang, S.; Lu, Y.; Li, L.; Shi, Y.; Lei, Y.; Yang, X.; Wu, Z. Effect of a hexacyclic triterpenic acid from Euscaphis japonica on the oleic acid induced HepG2 cellular model of non-alcoholic fatty liver disease. *Med. Chem. Res.* **2022**, *31*, 2209–2219. [CrossRef]
27. Lee, J.M.; Choi, S.S.; Park, M.H.; Jang, H.; Lee, Y.H.; Khim, K.W.; Oh, S.R.; Park, J.; Ryu, H.W.; Choi, J.H. Broussonetia papyrifera root bark extract exhibits anti-inflammatory effects on adipose tissue and improves insulin sensitivity potentially via AMPK activation. *Nutrients* **2020**, *12*, 773. [CrossRef] [PubMed]
28. Guru, A.; Issac, P.K.; Velayutham, M.; Saraswathi, N.; Arshad, A.; Arockiaraj, J. Molecular mechanism of down-regulating adipogenic transcription factors in 3T3-L1 adipocyte cells by bioactive anti-adipogenic compounds. *Mol. Biol. Rep.* **2021**, *48*, 743–761. [CrossRef] [PubMed]

MDPI AG
Grosspeteranlage 5
4052 Basel
Switzerland
Tel.: +41 61 683 77 34

Nutrients Editorial Office
E-mail: nutrients@mdpi.com
www.mdpi.com/journal/nutrients

www.ingramcontent.com/pod-product-compliance
Lightning Source LLC
Chambersburg PA
CBHW062100220926
51850CB00010B/2836
* 9 7 8 3 7 2 5 8 6 6 9 4 6 *